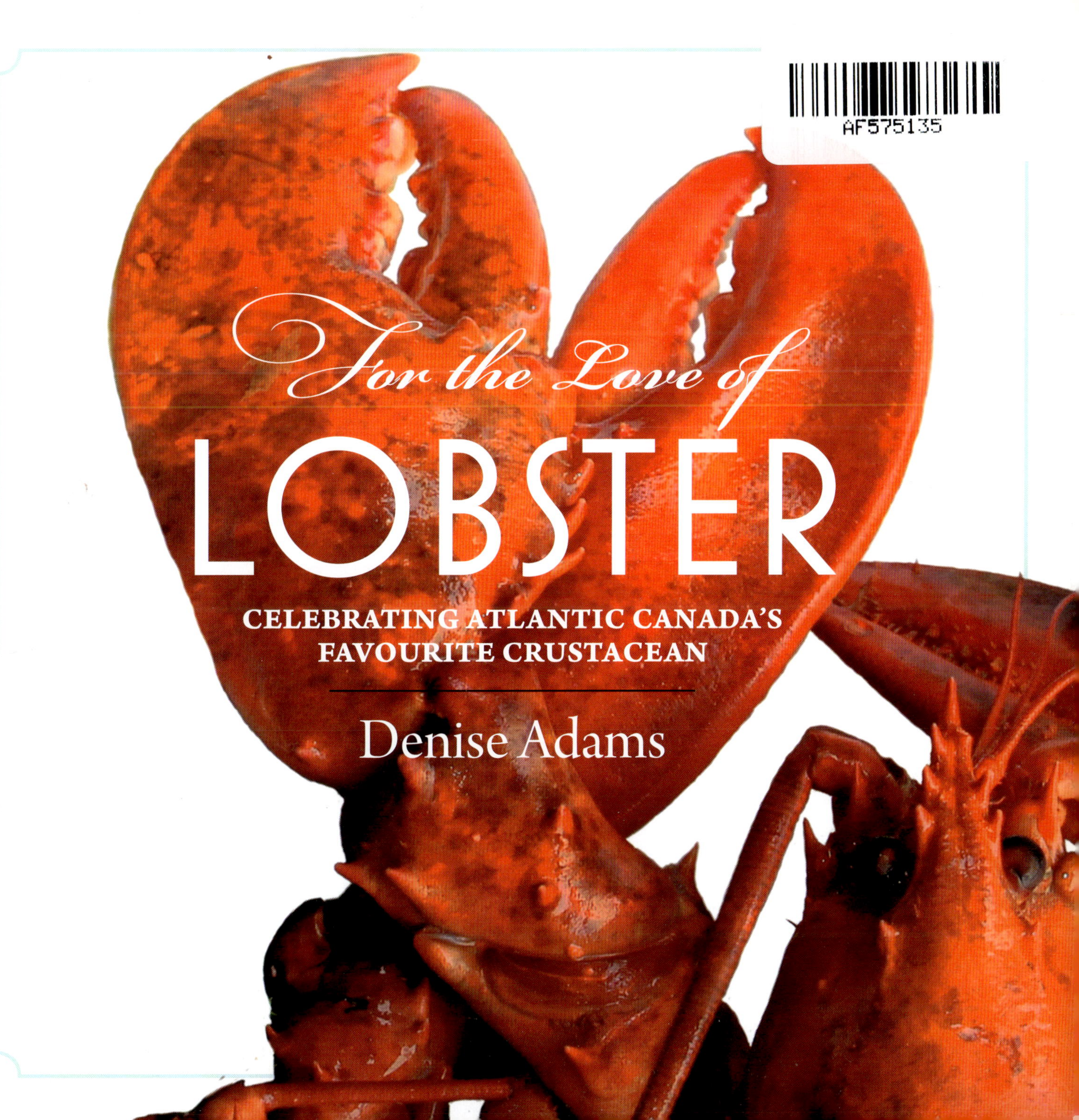

For the Love of
LOBSTER
CELEBRATING ATLANTIC CANADA'S
FAVOURITE CRUSTACEAN
Denise Adams

Nimbus Publishing Limited
3731 Mackintosh St, Halifax, NS, B3K 5A5
(902) 455-4286 nimbus.ca

Printed and bound in Canada

NB1237

Cover photo: Denise Adams
Design: Jenn Embree

Library and Archives Canada Cataloguing in Publication

Adams, Denise, 1961–, author
For the love of lobster : celebrating Atlantic Canada's favourite crustacean / Denise Adams.
ISBN 978-1-77108-398-0 (paperback)

1. Lobsters—Atlantic Provinces—Pictorial works. 2. Lobsters—Atlantic Provinces. I. Title.

QL444.M33A33 2016
595.3'84
C2015-908195-5

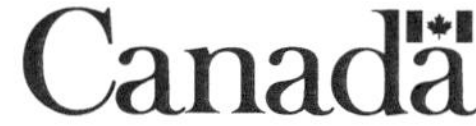

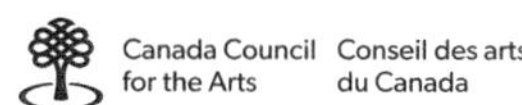

Nimbus Publishing acknowledges the financial support for its publishing activities from the Government of Canada through the Canada Book Fund (CBF) and the Canada Council for the Arts, and from the Province of Nova Scotia. We are pleased to work in partnership with the Province of Nova Scotia to develop and promote our creative industries for the benefit of all Nova Scotians.

CONTENTS

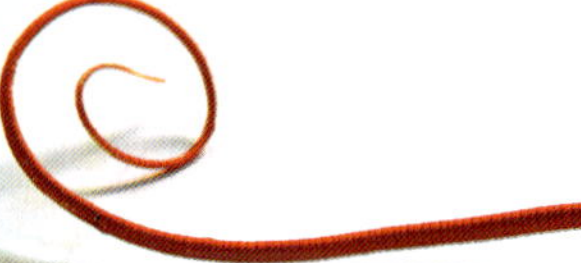

PREFACE

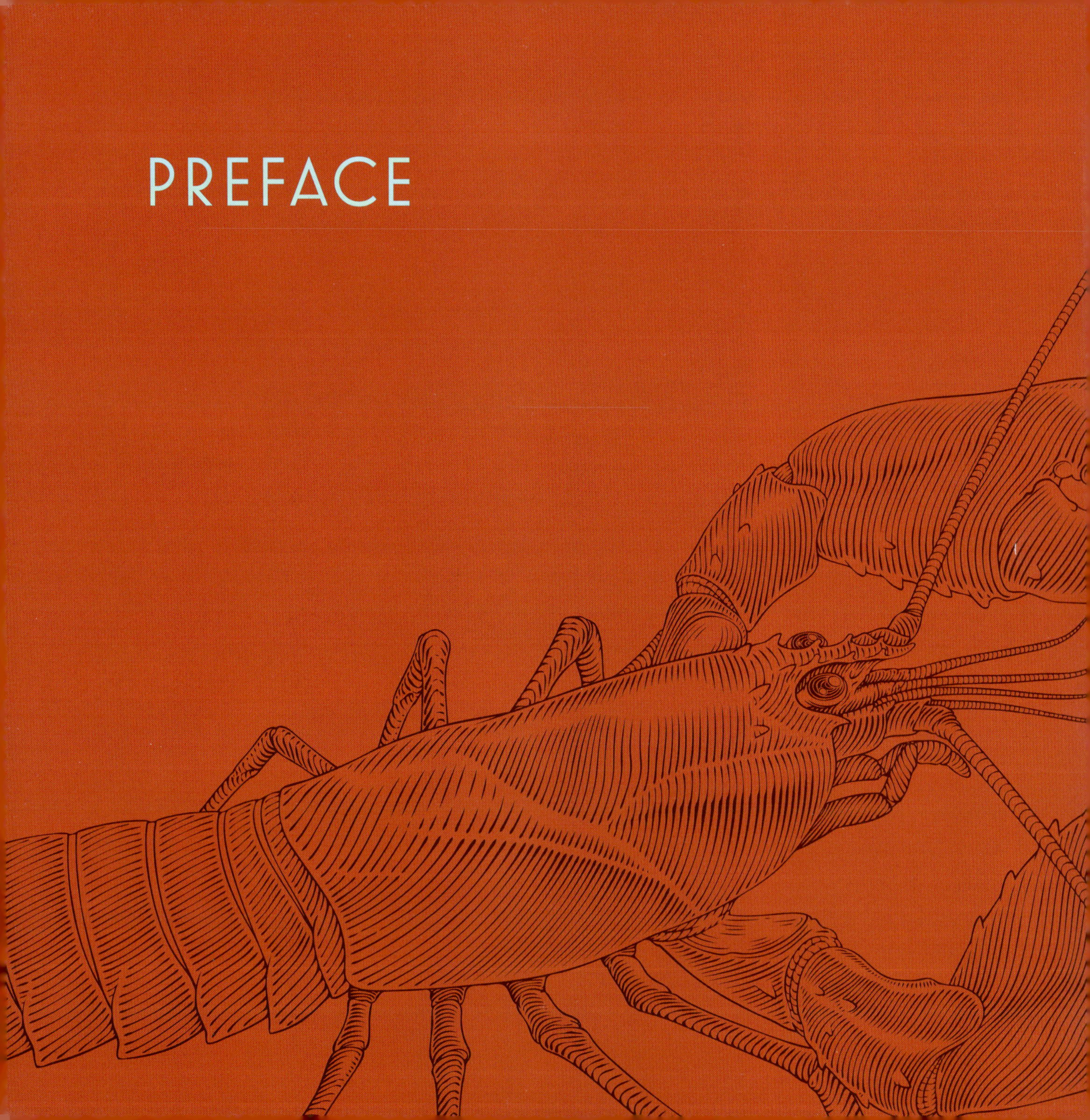

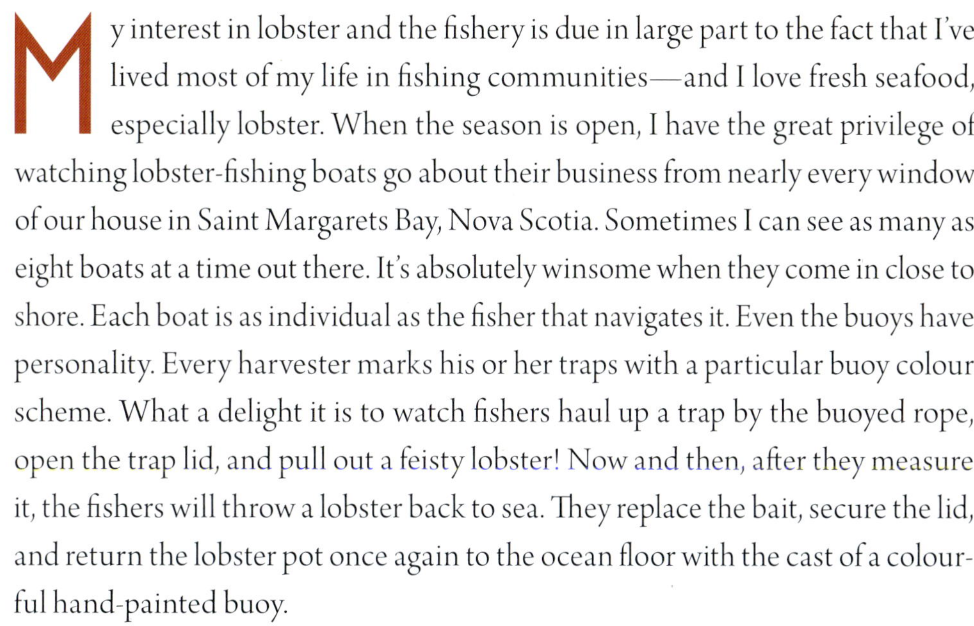

My interest in lobster and the fishery is due in large part to the fact that I've lived most of my life in fishing communities—and I love fresh seafood, especially lobster. When the season is open, I have the great privilege of watching lobster-fishing boats go about their business from nearly every window of our house in Saint Margarets Bay, Nova Scotia. Sometimes I can see as many as eight boats at a time out there. It's absolutely winsome when they come in close to shore. Each boat is as individual as the fisher that navigates it. Even the buoys have personality. Every harvester marks his or her traps with a particular buoy colour scheme. What a delight it is to watch fishers haul up a trap by the buoyed rope, open the trap lid, and pull out a feisty lobster! Now and then, after they measure it, the fishers will throw a lobster back to sea. They replace the bait, secure the lid, and return the lobster pot once again to the ocean floor with the cast of a colourful hand-painted buoy.

Atlantic Canada is home to the most sustainable and well-managed fishery around. I can enjoy watching the harvest worry-free. I've never been much of a meat eater, but I am an enthusiastic consumer of nutritious free-range lobster because I can see *how* and *where* it is harvested. Nothing beats the pristine, crystal clear, cold waters of the Atlantic Ocean. The ocean, its inhabitants, and all the activity it attracts is superbly photogenic. One doesn't have to be a professional to get great pictures. I knew from the start that this subject had the potential to make a visually interesting book. To the layperson, the lobster fishery is a beautiful

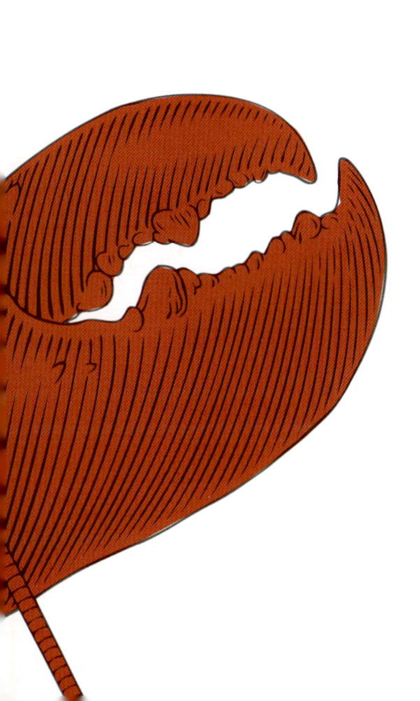

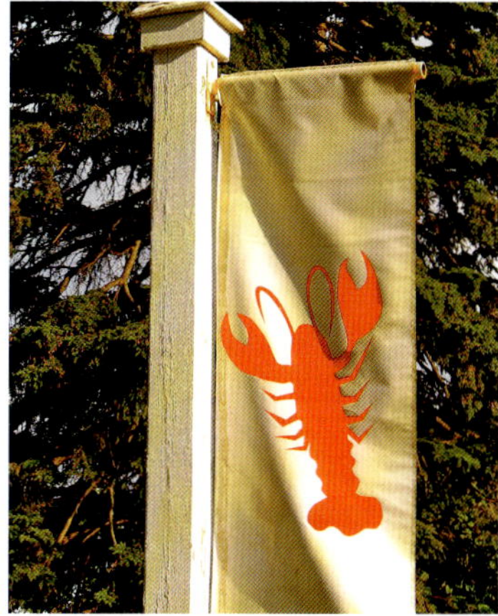

The lobster has become an emblem for our lifestyle in Atlantic Canada—and the freshest of fresh seafood.

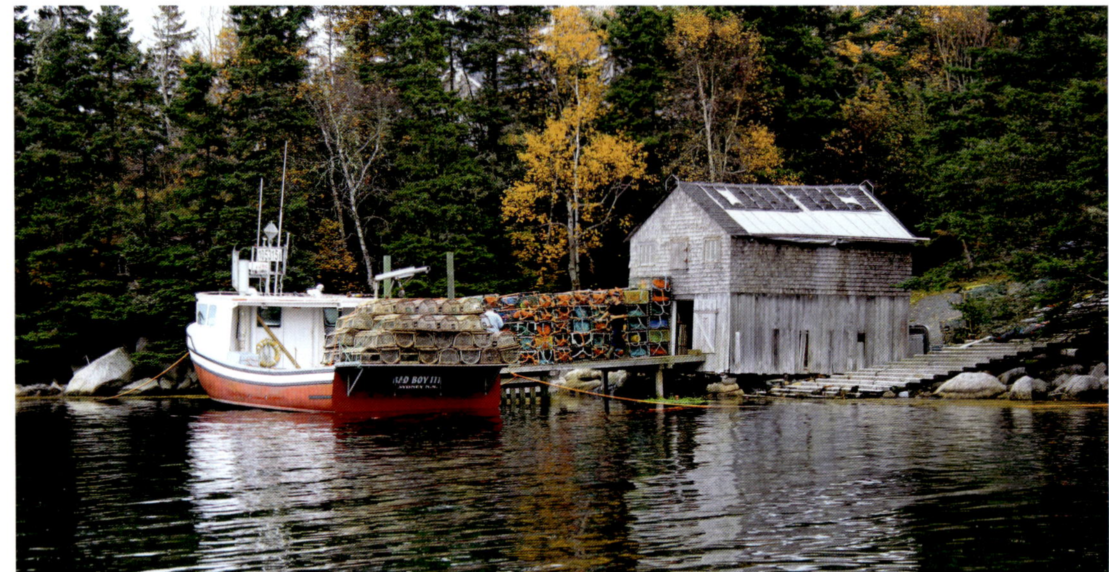

Boats and wharves are heavy with traps, a sure sign that the opening of the lobster fishing season is near.

sight, but harvesting lobster is a hard and risky job. There is so much mystery around it. I've always wanted to learn more about "lobstering" and the elusive crustacean. So I set out to write this book to share my findings with all who love and appreciate the ocean and its bounty as much as I do. There's so much more to lobster than the red-shelled delicacy on your plate.

Though I've always lived near the sea, my family roots are mostly agricultural. So it was with great humility that I took this topic on. I wished at times to have been the daughter of a lobster fisherman. Better yet, to be married to one. I must say, though, my husband, Rick, comes close enough. He's got Newfie blood and has always owned a boat. Rick has a keen eye for all things nautical: he can spot a wharf, boats, and fishing shacks in his periphery anywhere he goes. Rick was instrumental in helping me find great photo opportunities and finding people with good lobster-fishing stories to tell.

These are hard-working folks who reliably bring in a good catch despite the risk to their lives. I wrote about the lobster fishery as a bystander. You won't find anything about my

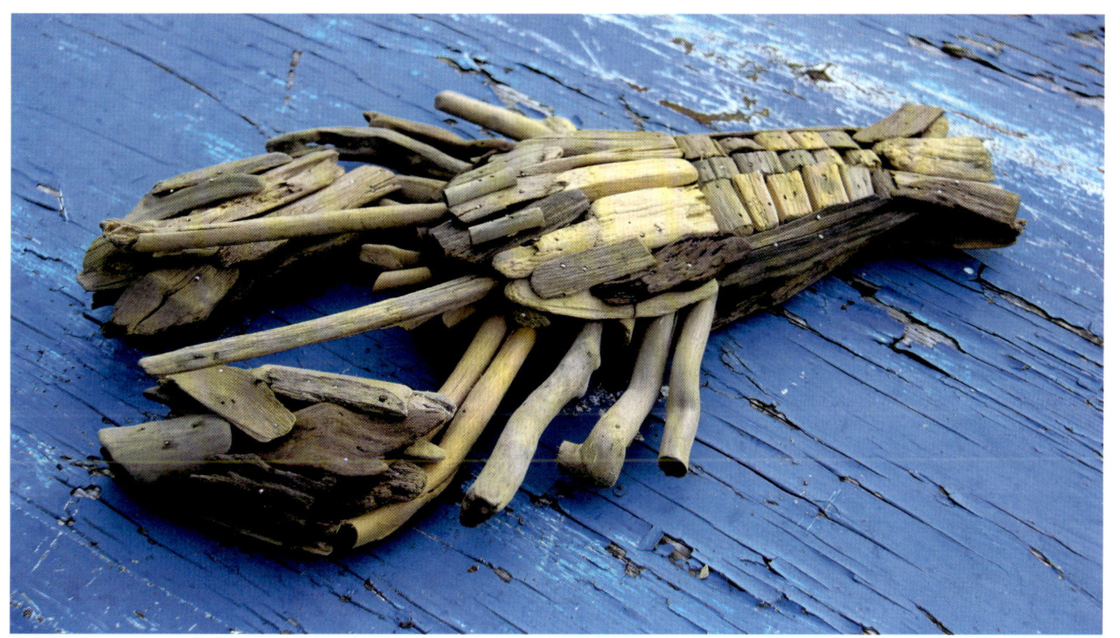

The lobster is celebrated in many forms of arts and crafts. This lobster is entirely made of driftwood collected from the same shores where lobster is caught.

adventures in the next pages. I give lobster harvesters, lobster pound handlers, lobster venders, and the humble lobster itself centre stage throughout the book. This is about *them*. It surprised me that some fishers felt they had nothing to say. They weren't convinced their work could possibly be book-worthy. I soon realized that I'd have to somehow gain their trust. I did this by validating their unique skills and knowledge, and by telling them that what they do is fascinating and definitely of interest to the public. Sometimes it took several visits. Timing was important. It was always easier after they'd had a beer or two at day's end. It seemed to me that lobster harvesters were either book-shy or flattered to be mentioned in one. A certain fisherman with a great story said: "I've never read a book in my life, so I'm not fussy about being in one. You can write my story but don't put me in it."

With all due respect, where I didn't have permission to quote someone directly, I tried my best to paraphrase. Equally important for me was to work hard at capturing how fishers

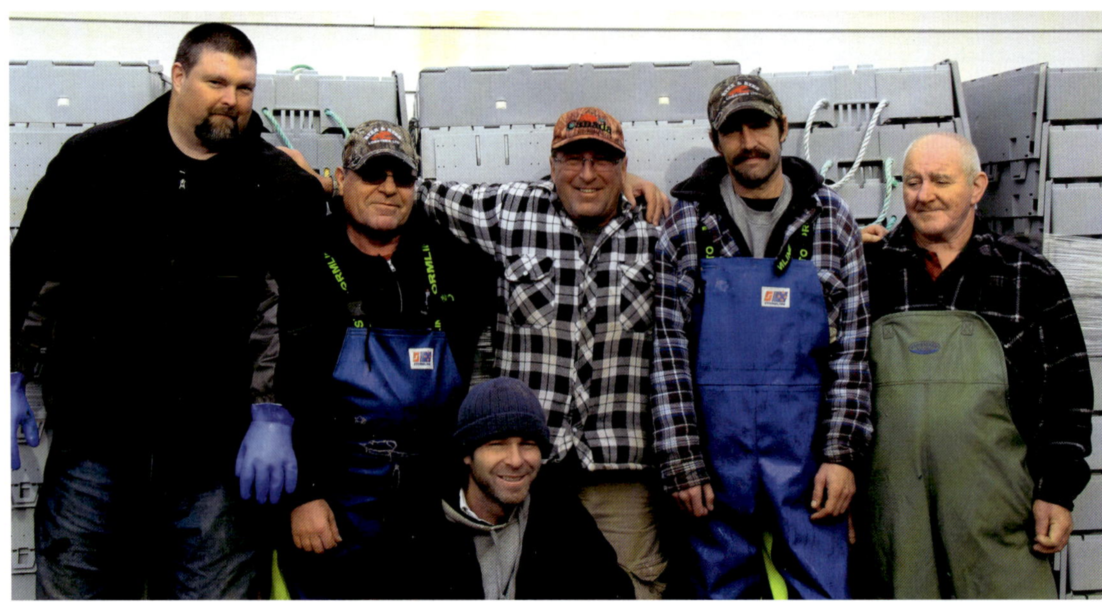

Some of the familiar faces at Ryers Lobster Retail: (L–R) Blair Eisner, Brian Fralick, Arthur Herritt, Mike Richardson, Harold Morash, and Adam Stoddard (bottom).

are so incredibly in tune with the rhythms of their watery surroundings. They are as one with the sea as the air they breathe. They have learned their trade by *immersing* themselves in it completely—from a lifetime of seeing and doing what their forefathers did. Their knowledge of the ocean commands much respect. Furthermore, lobster harvesters of the Atlantic provinces are brilliant entrepreneurs.

Most fishers I met, however, were happy and willing to let me photograph them while at work, to answer my questions, and to share a story or two. The salty jokes, tall tales, and true tales are what make this book such a fun read. And, oh my...the colourful boats, traps, ropes, and buoys on weathered ol' wharves were indispensible in making this project so visually rich. All the research I did aside, the best and juiciest discoveries about lobster and the life of fishers came from chatting with those who are directly involved in the fishery where I live, near Peggys Cove, Nova Scotia. The local fishermen and the workers at Ryers lobster

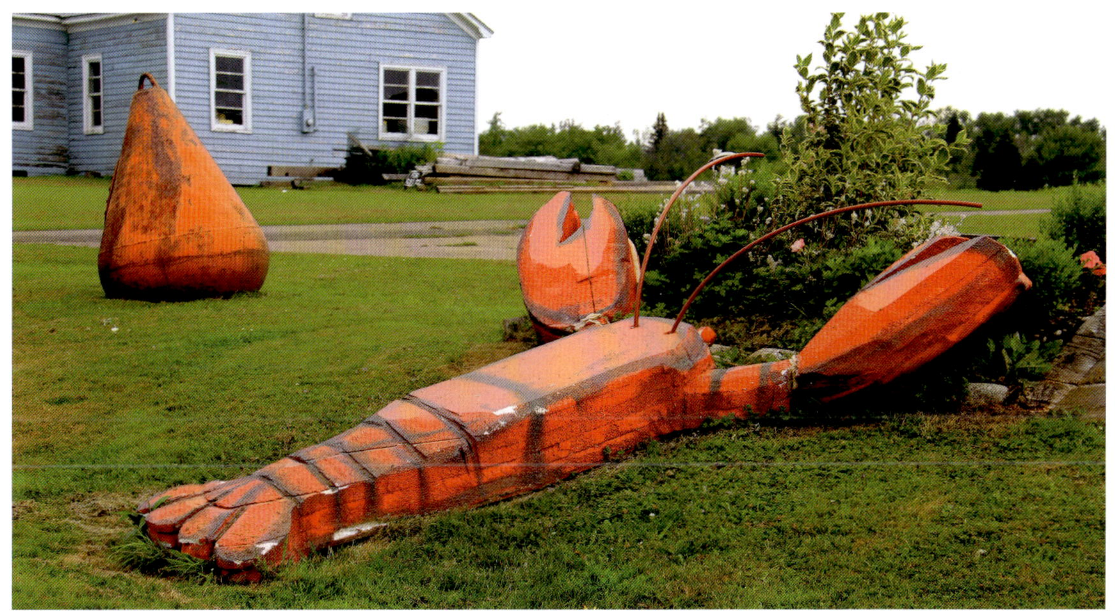

Lobster folk art is a common sight along the Atlantic coast.

pound in Indian Harbour were wonderful! Gratitude is especially owed to the Ryer family, in particular: Roger Ryer, his nephew Dean, daughter, Kim, and son, Kevin Ryer.

Ryer & Ryer Ltd Lobsters was established in 1971, just when the appetite for lobster started to peak, and has been going strong ever since. Many thanks also go to Dave and Joan Hoskin who run Ryers Lobster Retail roadside seafood store, where the ultimate no-fuss lobster experience happens on a picnic table overlooking St. Margarets Bay, where lobster is caught.

Thanks so much to everyone who participated for your generous and indispensible contributions about our favourite seafood: cold water–caught Atlantic lobster.

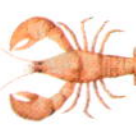

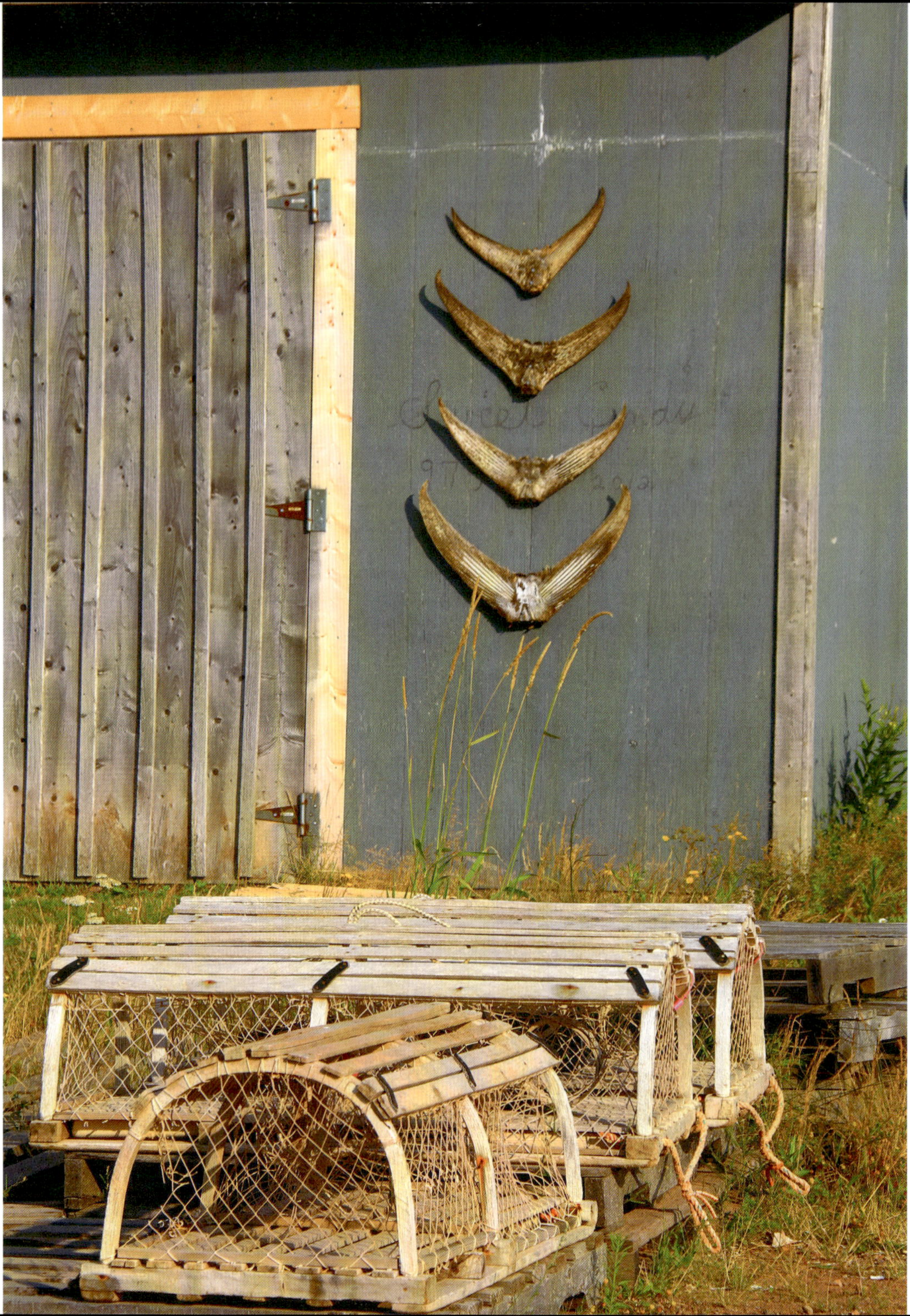

AUTHOR'S NOTE

The words "fisherman," "lobsterman," "fisher," "fisher-folk", "trapper," and "harvester" are used interchangeably throughout this book and are meant to include both the dedicated men and women involved in the lobster fishery. Those hardy gals who go out "lobstering" don't object to being called fishermen any more than they would object to lobsters still being weighed by the pound and measured by the inch. These terms have been used liberally throughout this book to reflect the spirit of what still takes place on the boats and wharves of the Atlantic provinces in the twenty-first century.

MEET *the* CREATURE

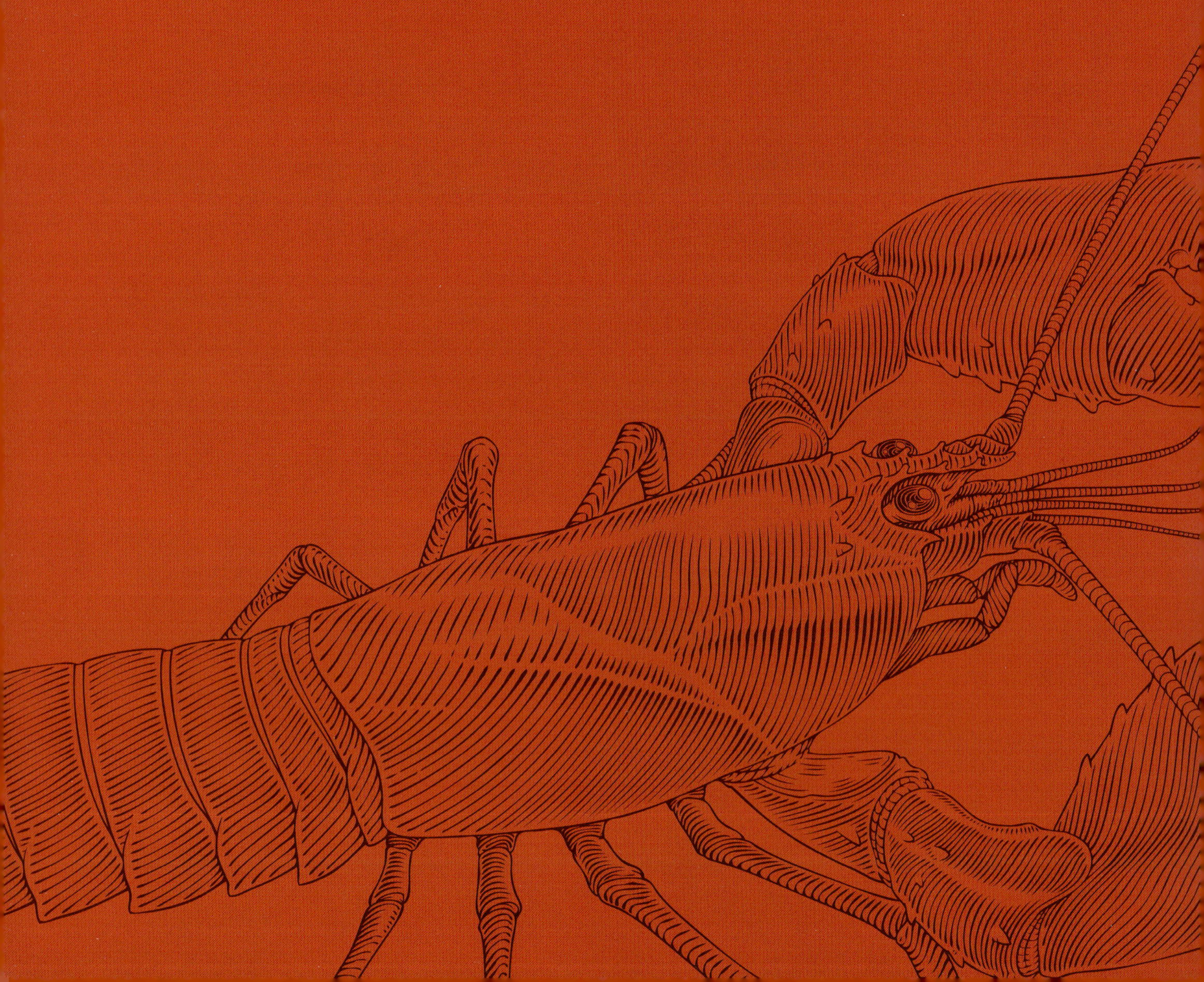

Introducing *Homarus americanus*

What could possibly be more Canadian than maple syrup, polar bears, and hockey? What on the east coast gets photographed and represented more than lighthouses, wharves, boats, and buoys? You don't have to go very far in Atlantic Canada to see an image of a red lobster on a roadside billboard or on a sign pointing to a simple little mom-and-pop seafood eatery. Lobster began its sharp rise to prominence as a food delicacy worldwide in the 1960s and today it is, in many ways, Canada's ambassador to the world. But as recognizable as the lobster is, how well do we really know this curious underwater creature?

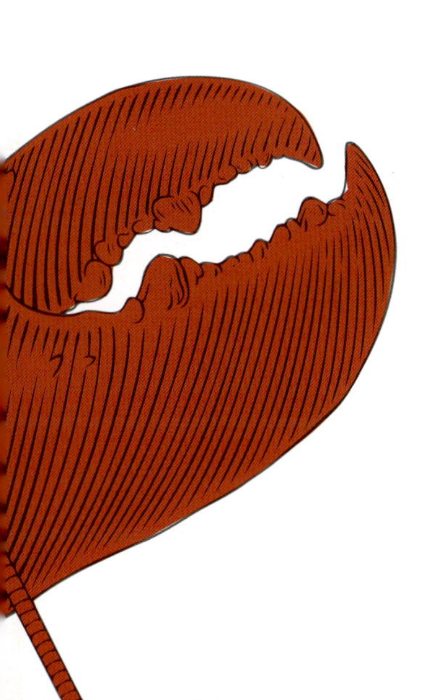

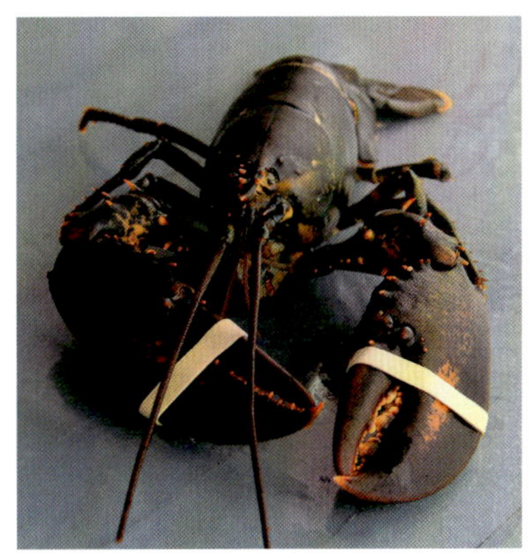

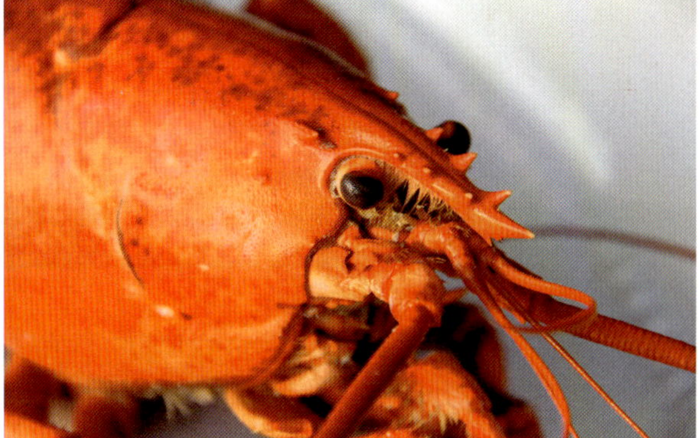

We immediately think of lobster as red, but a fresh live lobster (left) is predominantly black-green with only slight orange and blue markings. Only when lobsters are cooked do they turn red (right).

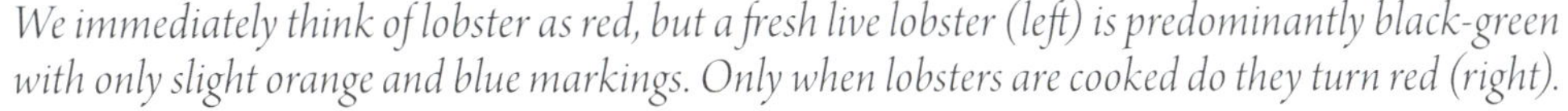

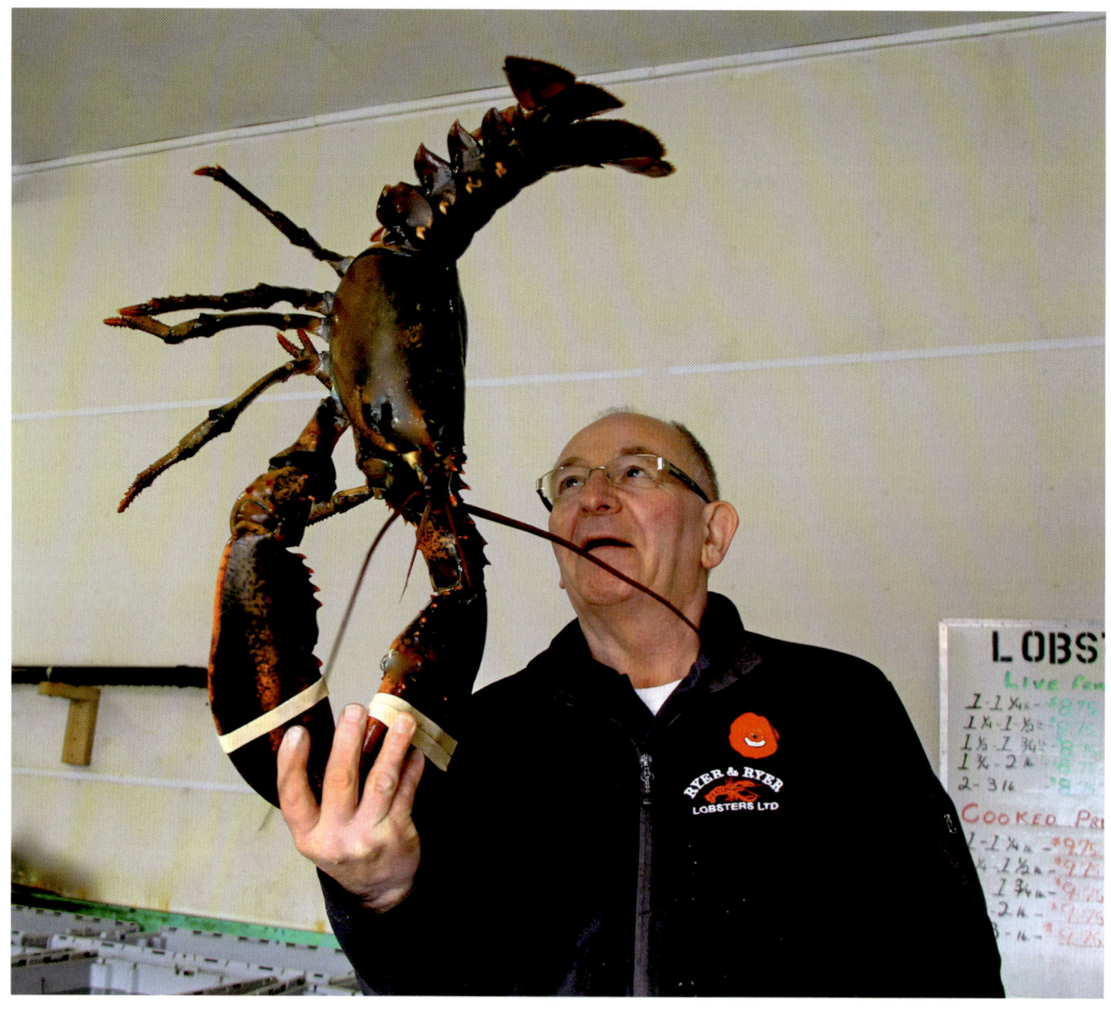

Lobster salesman Dave Hoskin shows off a live 8-pound lobster from Ryers Lobster Retail outlet in Indian Harbour, Nova Scotia.

QUIZ:
How Well Do You Know Lobster?

We all know lobsters live in the sea and turn red when cooked. But there is so much more to this mysterious aquatic creature than you might think. See how well you know Atlantic lobster by answering these twenty questions.

Part I: True or False

1. Two hundred years ago, lobsters were nearly three times larger than they are today.
2. Lobsters have claws when they first hatch.
3. Lobsters have a well-understood life expectancy.
4. Lobsters mate for life.
5. Early settlers to Atlantic Canada preferred fish over lobster.
6. A lobster can grow a new limb, like a claw or leg, when one is lost.
7. Lobsters have been known to eat each other.
8. Some lobsters are right-handed and others are left-handed.

Part II: Multiple Choice

9. How can you tell a female lobster from a male lobster?
   A. *By the size of the teeth on their claws*
   B. *By the length of their antennae*
   C. *By the number of segments on the tail*
   D. *By the shape of their swimmerets*
10. The biggest lobster on record was caught off the shore of
    A. *Maine, USA*
    B. *Nova Scotia*
    C. *Newfoundland*
    D. *Greenland*
11. Lobsters can live out of water for
    A. *20 minutes*
    B. *2 hours*
    C. *2 days*
    D. *2 weeks*
12. The greatest quantity of eggs a large female lobster can lay at one time is
    A. *50*
    B. *500*
    C. *5,000*
    D. *50,000*
13. Lobsters are most active
    A. *All the time*
    B. *In the morning*
    C. *At noon*
    D. *At night*
14. Lobsters use their long and short antennae
    A. *To smell and feel their way around*
    B. *To communicate with other lobsters*
    C. *To attract the opposite sex*
    D. *All of the above*
15. What part of the lobster is the tomalley (the green paste inside its body)?
    A. *The brain*
    B. *The liver*
    C. *The stomach*
    D. *The lobster's blood turns green when cooked*

Part III: Short Answer

16. What is a lobster's most feared predator?
17. Why do lobsters in warm southern waters not have claws?
18. Why are soft-shell lobsters less marketable than hard-shell lobsters?
19. Can lobsters swim, or do they just crawl?
20. Are there parts of the lobster that, if ingested, could make you sick?

(For answers, see page 138)

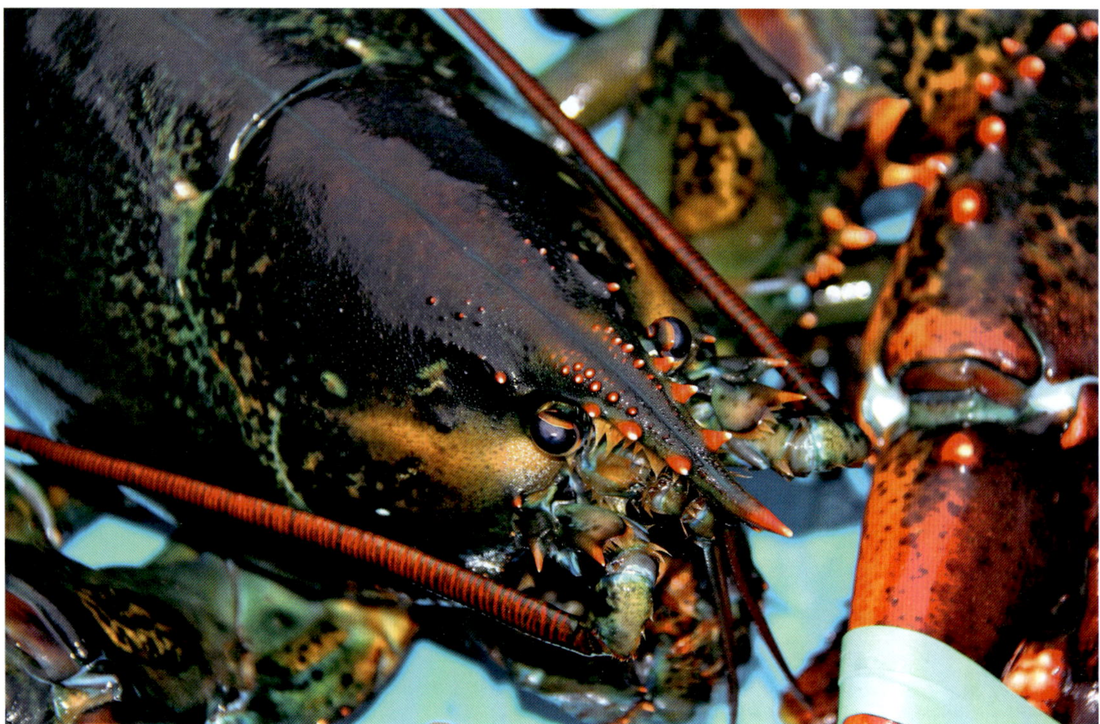

Lobsters have very insect-like faces for good reason.

We can all agree lobsters are not exactly cute and cuddly. But not only do they look like bugs, they technically *are* bugs: the lobster is an underwater sea insect. In fact, its early Latin name, *locusta,* is due to its resemblance to locust (grasshopper). The Old English name for the odd-looking animal was *loppestre,* which is believed to have derived from *loppe,* meaning spider. From there, the word morphed into *lopystre,* then *lapstar,* and was eventually standardized as the word we use today: lobster, which dates back only two centuries. The French call the crustacean *homard.* Its origin can be traced to lobster's early Greek name, *Homarous,* which is believed to have come from the Greek expression "made for Homarh," which refers to the ancient Greek poet "Homer," who apparently acquired a taste for the crustacean.

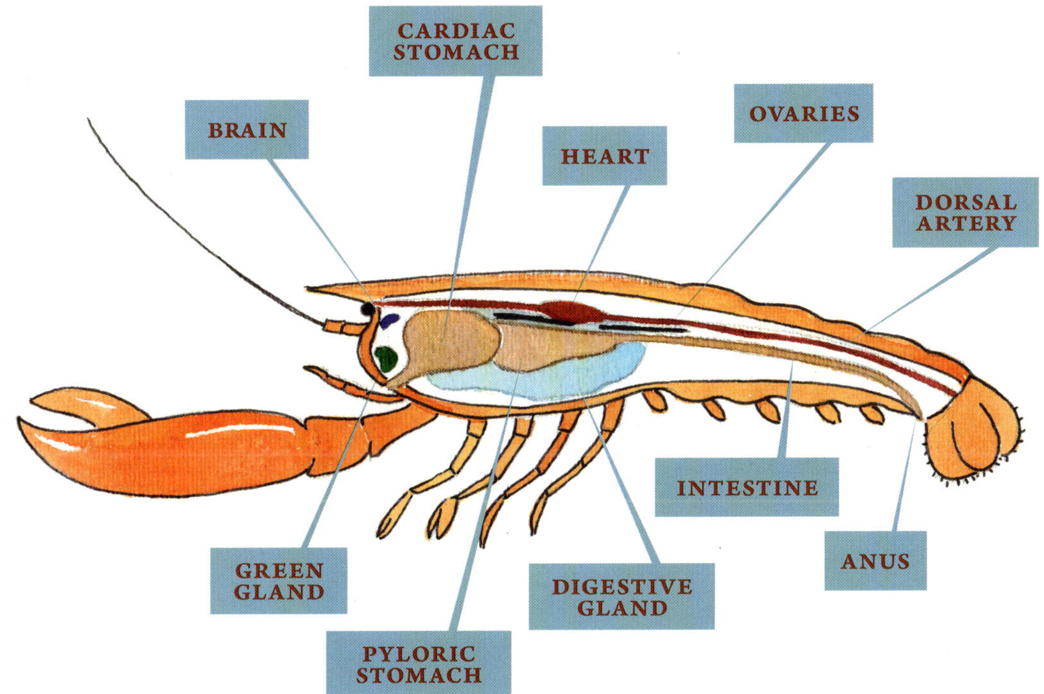

Lobster cross-section of adult female.

Lobsters, along with spiders, cockroaches, and scorpions, belong to a group of insects called arthropods. These are invertebrates (no spine, no bones) with segmented bodies incased in an exoskeleton. More than 80 per cent of all land and aquatic insects are arthropods. Found all around the world, they range in size from the miniscule head lice to the common grasshopper to the 8- to 9-centimetre giant Asian praying mantis. What makes the lobster an unusual arthropod compared to other "big bugs" is its potential to grow to enormous lengths. Furthermore, no matter how large a lobster becomes, its brain remains no bigger than a grasshopper's!

Humans usually find insects, especially large ones that bite, repulsive. A plate-size lobster capable of wielding a very nasty bite with those spiky sharp claws is, for curious reasons, an exception to the rule. Why? The answer, partly, is the colour of cooked lobster: *red*. Several

Our Lady of the Lobster

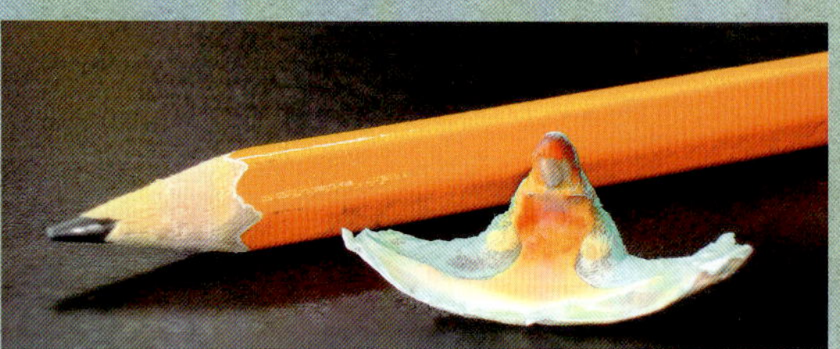

A highly revered deity in the Roman Catholic faith is Mary, mother of Jesus. There is a little bony structure hidden inside the lobster, just behind the eyes, that looks remarkably like a typical statuette of the Virgin Mary. This was felt among some Catholic fishermen to be a sign of her motherly protection over them when at sea. Not much bigger than the tip of a pencil, this mythical-looking structure is part of the lobster's digestive system. It is a tiny central tooth called an ossicle that is part of the gastric mill in the lobster's foregut. It is used in conjunction with denticles, or calcified primitive teeth, to grind food. Together, they function much like stones in the gizzard of birds.

studies conducted on the psychological effects of red prove the colour stimulates appetite: it increases pulse, heart rate, and blood pressure, which increase our metabolism, and in turn heightens appetite. Also, once cooked, the lobster's colour paired with its fresh-from-the-sea fragrance and the delicious flesh inside its shell get us past the *ick* factor in no time. This wasn't always the case, though.

There are many turn-of-the-century references to beaches at low tide along the North Atlantic Sea Board, including those in the Atlantic provinces, crawling with lobsters. As recently as a hundred years ago, lobsters were perceived to be about as valuable and visually appealing as nasty, biting sea insects. Not much attention was paid to lobster as a food source. They were reviled. In fact, lobsters are reported to have been collected mostly as fertilizer, their nutrients added to soil for growing crops. They were only eaten if food became severely scarce.

Something has changed.

How, in such a short period of time, did lobster go from having such a bad reputation to becoming the luxury food of the twenty-first century? We'll discuss the rise of lobster later on (see p. 47). First, let's learn a bit more about the creature itself.

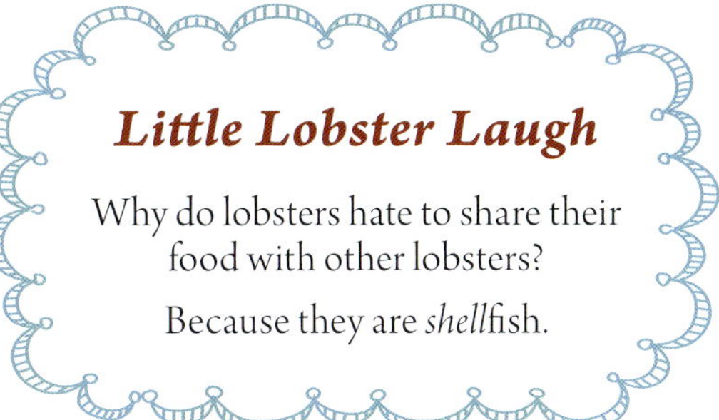

The Elusive American Lobster

The new-found high value of lobster has prompted numerous studies of the crustacean. We know it is most active at night. We know it prefers a diet of fresh animal protein, like mussels,

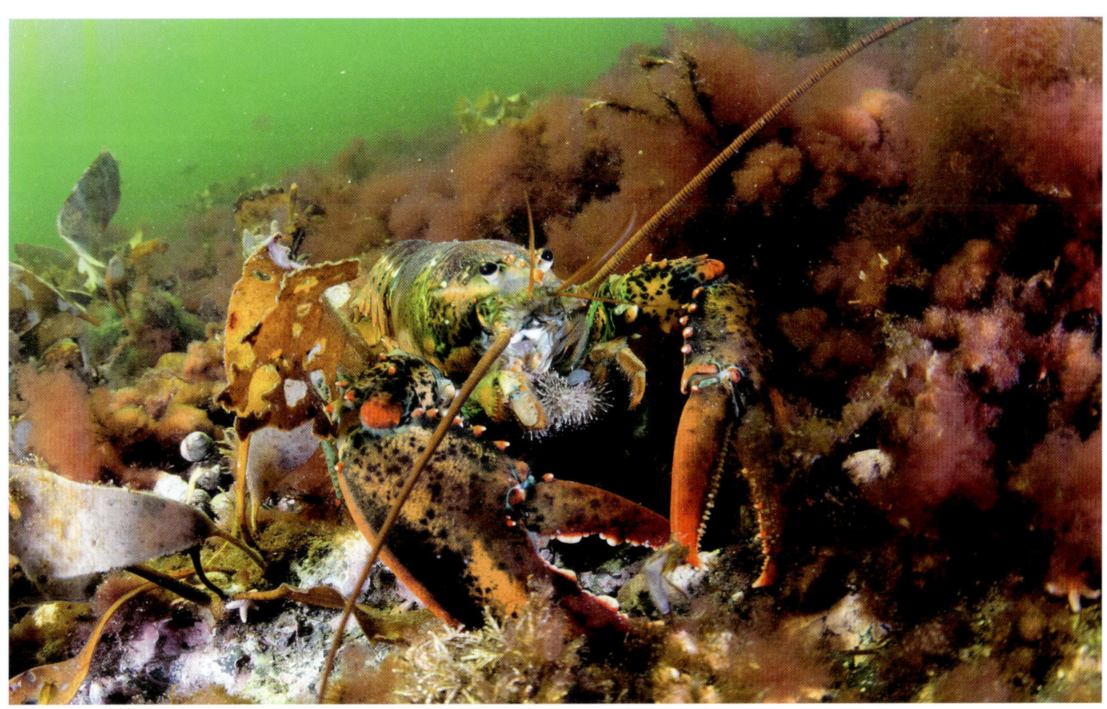

American lobster, Bonne Bay, Newfoundland.

Lobster Diversity

Lobster caught in the cold waters of the Atlantic Ocean are called "American lobster" (*Homarous americanus*). It has many relatives throughout the world's oceans and even in some fresh bodies of water (crayfish). There are basically two types of saltwater lobster: those with claws and those without. Here are a few of the most interesting saltwater lobster species found throughout the world.

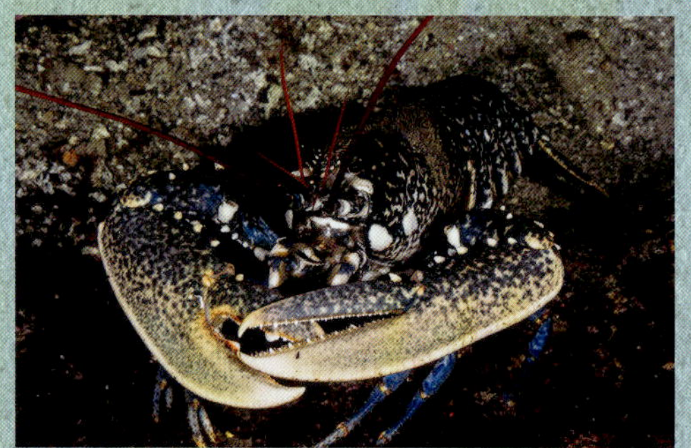

European lobster

This is the closest relative of the American lobster; the two are almost identical. European lobster live in waters of the northeastern Atlantic Ocean, have smaller claws, and are more prone to blue-green markings. Their range spreads from Norway to Morocco, with an annual harvest of two to three thousand tons. (This is small in comparison to the sixty thousand tons of American lobster landings per year.)

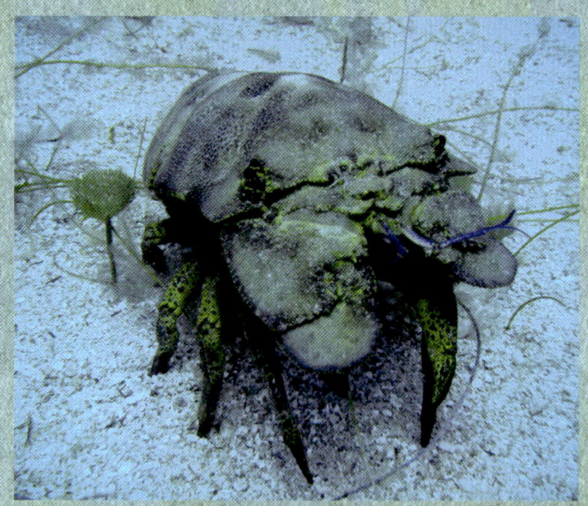

Sculptured slipper lobster

There are several species of slipper lobster widely distributed throughout the world's warmest oceans. They don't have true claws: what appear to be flat claws are actually enlarged antennae that also serve as "shovels" for foraging and hiding in sand. The sculptured slipper lobster inhabits the waters of Florida to northeast Brazil and the Pacific Ocean near Hawaii and Polynesia. A mature slipper lobster grows to only 20 centimetres long, with the edible tail taking up about half that length. It is too small and too spread out geographically for a viable commercial fishery, but Caribbean Islands locals enjoy its tender flesh and flavour—when they can catch it.

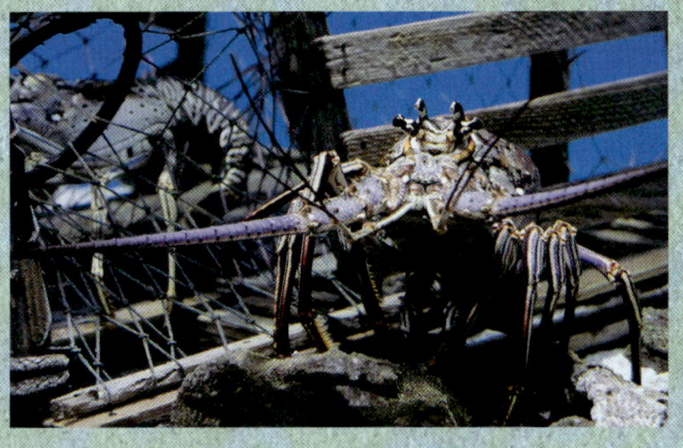

Spiny lobster

This clawless crustacean, also called "rock lobster," is in fact more closely related to shrimp. It inhabits the tropical and subtropical waters of the Atlantic Ocean from Florida down to the Caribbean Sea and the Gulf of Mexico. The spiny lobster is robust, with two folded-back spiky antennae that are as long as its entire body. Its average length is about 20 centimetres but can reach up to 45 centimetres. Although the spiny lobster's flesh (tail) is tougher and less flavorful than that of the American lobster, some populations suffer from overfishing. Now that it has recently been farmed successfully, the hope is that natural stocks will get a chance to rebound. The spiny lobster fishery is the largest commercial fishery in Florida at 6 million pounds a year, and is worth $20 million annually.

Furry lobster

The furry lobster, also dubbed the "Yeti crab" for its whitish, hairy appearance, was discovered in 2005 around deep-sea hydrothermal vents of the South Pacific Ocean near Easter Island. As the name suggests, the furry lobster is covered in sinuous hairs, especially the pincers. This 15-centimetre-long crustacean is completely blind. It lives in waters that are inaccessible and lethal to most ocean species. Because of the depths and toxicity of the waters in which it lives, the furry lobster is not viable for consumption.

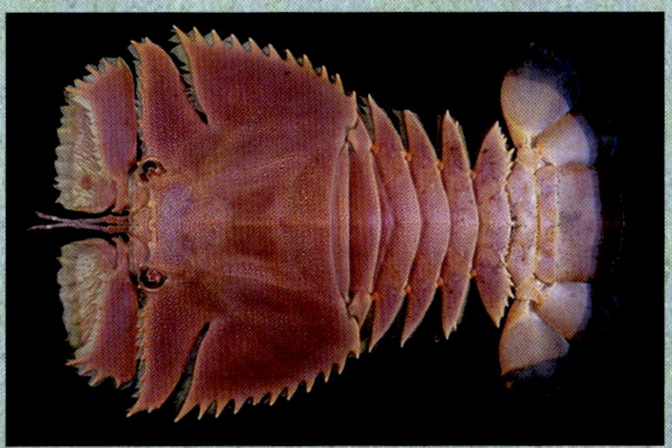

Japanese fan lobster

This flat, clawless lobster is similar to the slipper lobster and measures approximately 23 centimetres. The contour of its body is almost entirely covered in sharp, curved spikes. It is largely landed as a bycatch by trawlers of other fisheries. Inshore shellfish divers also harvest them when they are found, but there is no active fan lobster fishery. Although a popular crustacean in fish markets from the Philippines to the Korean Peninsula and Japan, there is little data on quantities being caught.

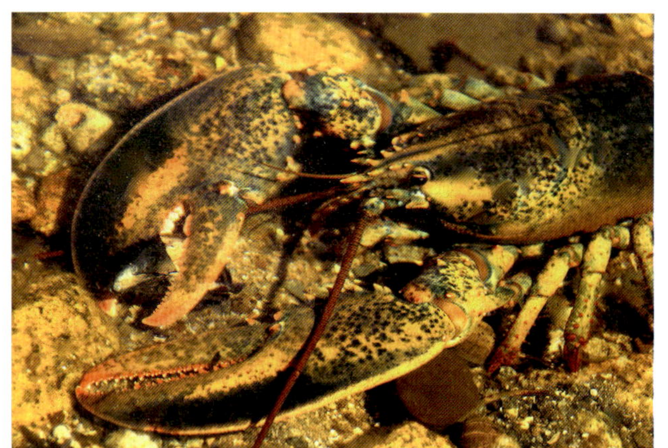

The older a lobster gets, the easier life gets. This massive 8-pound lobster is crushing a mussel for its meal.

clams, sea urchins, small crabs, and even juvenile lobster! They also nibble on sea plants. However, their strong sense of smell, mostly from their small antennae and feet ends, will lure them into a baited trap of dead fish for a quick, easy snack. We know they must moult (shed their shell) several times a year in order to grow, and less so as they age. More recently, we have learned how they procreate (a topic we'll explore in the next chapter). We know so much more than ever before, yet still so little.

Out of sight in their dark, briny, liquid world, lobsters can only be observed closely in captivity. Even then, they remain elusive. When studied in large aquarium tanks embellished with all of the natural materials a lobster would have on the sea floor, there is no way of knowing if their behaviors are representative of what goes on in the wild. Very little is known about how they interact with other creatures and with their own kind.

Fishers are mystified when a spot that had no lobsters entering traps one year will yield them reliably the next. Why is one kind of bait a better lure here than there? How many lobsters in a trap's vicinity don't find their way in? Lobster fisher Jean Belliveau of Belliveau Cove, Nova Scotia, sums it up: "Where to set your lobster pots is all guesswork. You don't know what goes on down there."

Many questions remain for lobster researchers as well. Why do lobsters frequent some spots of the ocean floor more than others despite geological similarities? How wide is their range, or are they constantly on the move? Where do females go to spawn? How long will a lobster stay in a trap before it seeks escape? These are just a few questions that would benefit the fishery if answered. But there are limitations to studying lobsters in the wild. They often live in low-light areas and are most active at night, so attaching a camera to a lobster would be futile, and human presence (diving) affects lobster behaviour.

Lobster Longevity

What is most baffling about lobsters is, unlike we humans and other animals, they don't seem to deteriorate with age. Unless a lobster succumbs to predation or disease or lands in a trap, it seems to keep going and going. There is actually a scientific term for indefinite life expectancy: "negligible senescence." Like a fine wine, a lobster's quality of life appears to improve with time—not so much its flavour or texture, though. A lobster's meat becomes too fibrous once

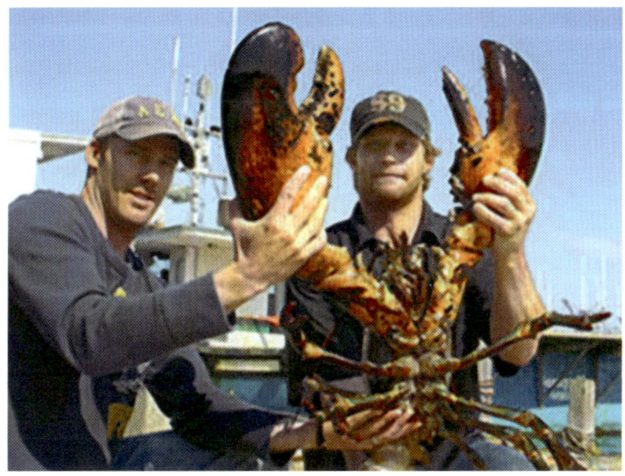

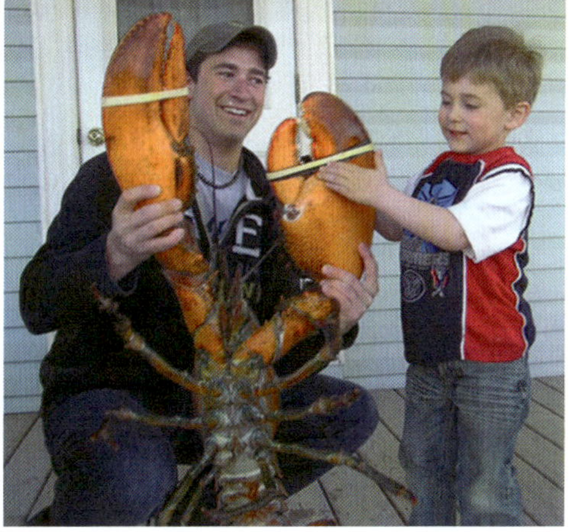

(Above) Julien German (left of lobster) caught a lobster of prehistoric proportions, weighing 23.2 pounds, off the coast of Meteghan, Nova Scotia, mid-May 2010. To put its size in perspective, photographer Marcel Cottreau had his four-year-old grandson (right) stand next to the crustacean.

New claws are red and gelatinous when they first appear.

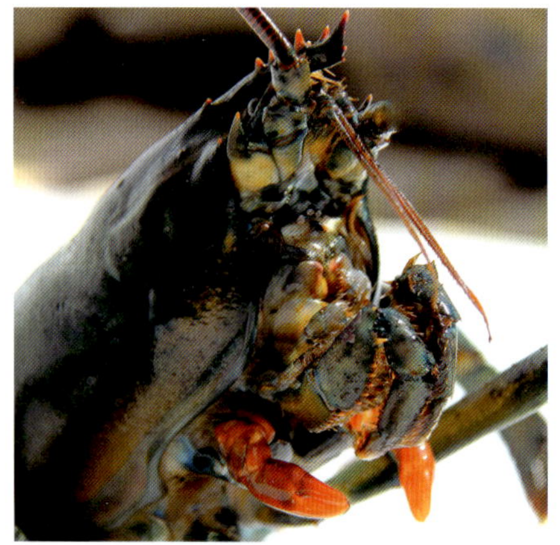

A new claw can only grow with each moult.

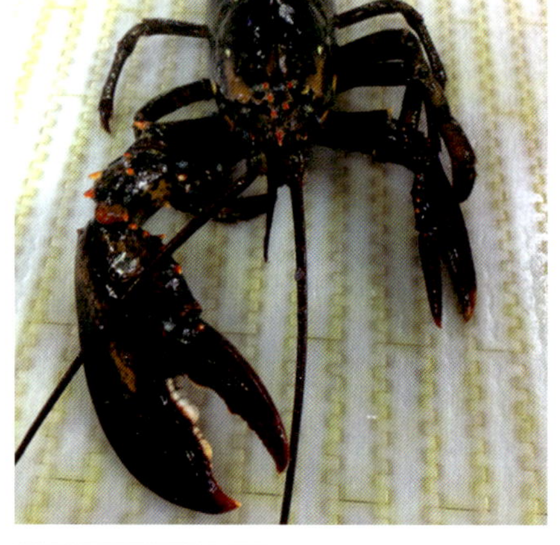

This market-size lobster shows two new claws. It will take up to 6-7 years to grow a claw proportionate to its body size.

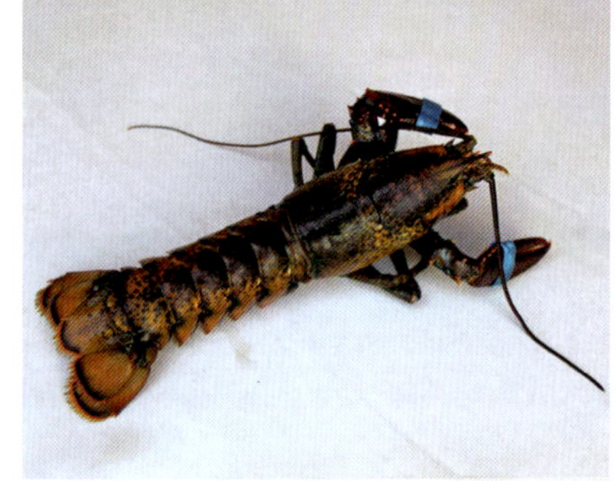

it weighs over 3 pounds, around age twenty. It takes five to seven years (depending on water temperature) for a lobster to reach legal market size. A 3-pounder would be fifteen to twenty years old. A 20-pound lobster would be seventy-five to one hundred years old. The only time a lobster is frail and vulnerable is in the week following a moult (shedding of the old shell).

So how long can lobster live? The simple answer is: as long as they are being caught, we can't know for sure. There is no surefire way to determine a lobster's age. It's not like counting the rings of a tree trunk or core sediment layers of a glacier. On average, though, landed lobsters weigh 1–2 pounds. The largest lobster ever recorded, according to the Guinness Book of World Records, was caught off the shores of southwestern Nova Scotia in 1977. It weighed 44 pounds and was 1 metre long. It is said it had claws powerful enough to snap a man's arm off. The estimated 150-year-old lobster was eventually sold to a high-end New York City restaurant for display in a large aquarium.

Longevity is definitely an advantage for lobster. It can grow to a size that makes getting stuck in a baited trap impossible. Instead, an unusually large lobster is most often discovered as bycatch, getting tangled in fishing gear. Also in lobster's favour is that the older

Lobster and the Key to Eternal Life

Recently, scientists discovered that lobsters have something called telomerase that protects their DNA. The tip of every lobster chromosome ends with a chemical cap called a telomere; these act like the plastic tips at the end of shoelaces, and prevent lobster DNA strands from fraying. Research has shown that cancerous cells divide uncontrollably and act as though immortal because they are exposed to telomerase. Drugs that deny telomerase to cancerous parts of the body are still in clinical trials. Some scientists theorize that in the future, ways of using telomerase could prevent normal cell death and possibly extend human life to unprecedented longevity.

and larger it gets, the harder and spinier it becomes, making it difficult for predators to access the meal inside its armour.

In its senior years, everything seems to get easier for lobster. There is no mid-life crisis, no over-the-hill downward spiral, no senior moments of forgetfulness, no need for Viagra or hormone-replacement therapy to encourage sexual appetite...no problem getting old. To top it off, lobster experience no muscle loss and can even regenerate lost limbs. It should be no wonder that scientists are looking to lobster for organ regrowth, the secret to longevity, and even eternal life.

Aging female lobster experience no menopause, no osteoporosis, and no wrinkles. Instead, the older a she-lobster gets, the more eggs she produces and, as a result, the more offspring make it to maturity. It is difficult to determine when a female lobster is sexually mature. It seems to be completely dependent on environmental conditions. In Nova Scotia's waters, for example, female lobster as small as 4 inches have been found with eggs attached to their abdomen: an indication that copulation has taken place. A 1-pound female is about seven years old; she lays approximately 8,000 eggs. Female lobster weighing over 3 pounds,

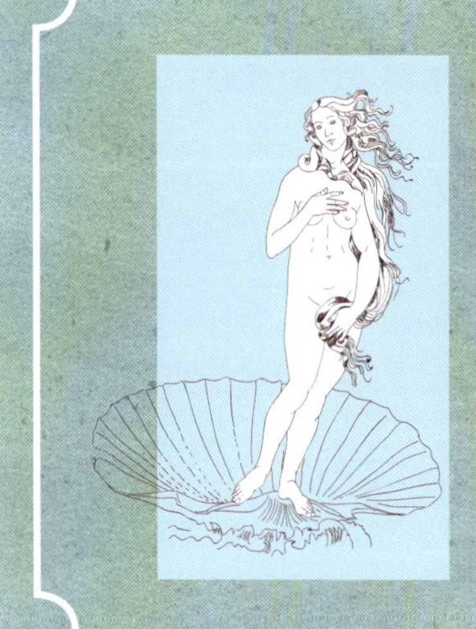

Lobster Love Pill

The term "aphrodisiac" can be traced back to ancient Greek mythology. The diety *Aphrodisios,* "pertaining to Aphrodite," is the Greek goddess of fertility, love, and beauty. Born of the sea, Aphrodite had ocean creatures as playthings in her "love games." Thus, sea creatures developed a reputation for enhancing sexual appetite. This reputation endures to this day. In his 1968 book, *The Virility Diet,* Dr. George Belham recommends lobster among other foods to remain sexually active. Today, we know that zinc and B-12–rich lobster meat are both necessary nutrients in maintaining sexual desire.

fifteen to twenty years old, can put out the same amount of offspring as ten younger, canner-sized (smaller, legal-sized) female lobster put together: about 50–80,000 thousand eggs! A 9-pound female, fifty to sixty years old, produces as many as 100,000 eggs. The reputation of lobster being an aphrodisiac predates any of these findings. There must be something to it!

In the next chapter we'll explore more fascinating facts about lobster, including their elaborate courtship rituals, the surprising connection between the female's moult and successful reproduction, and the lobster's life cycle.

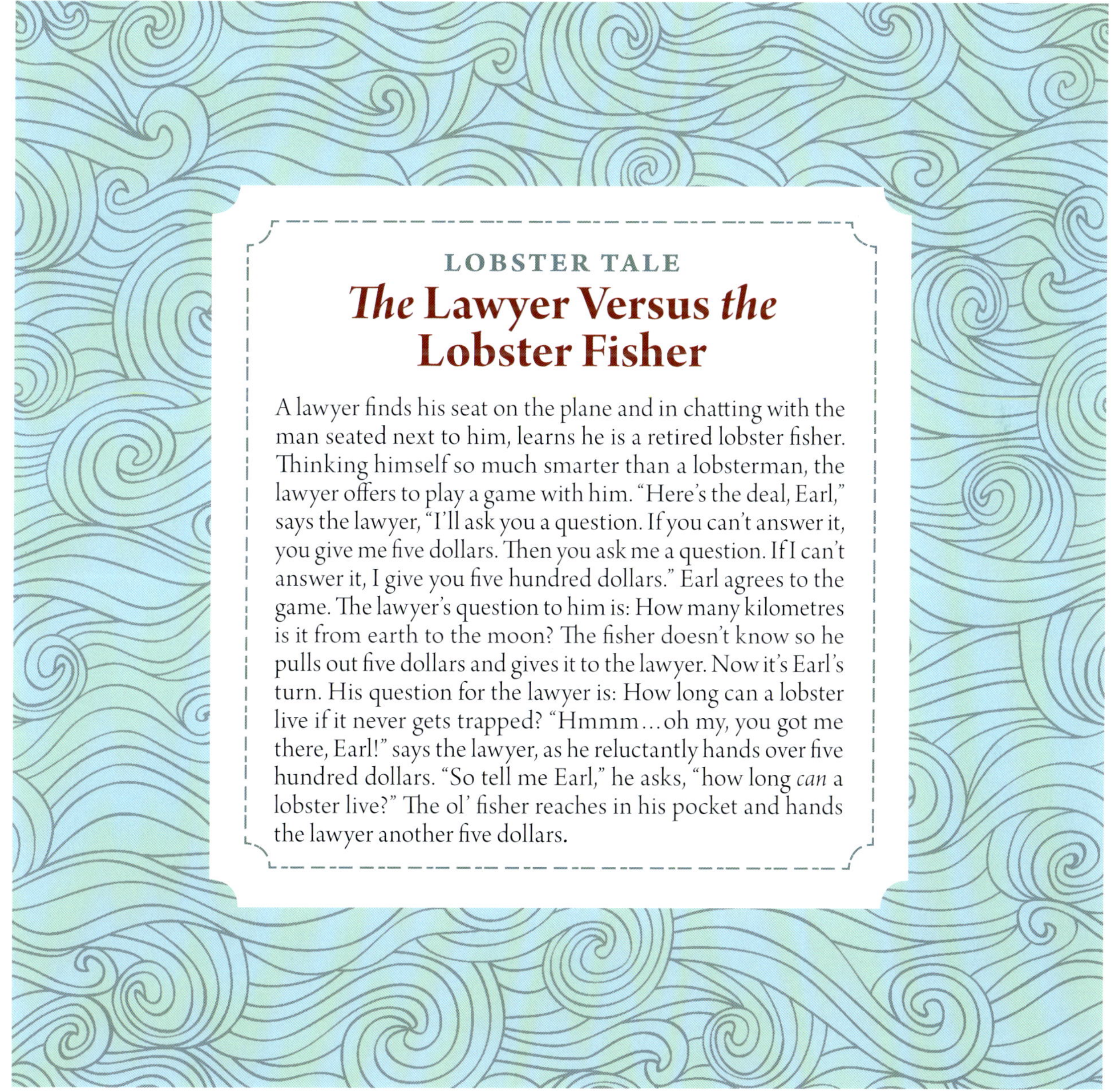

LOBSTER TALE

The Lawyer Versus *the* Lobster Fisher

A lawyer finds his seat on the plane and in chatting with the man seated next to him, learns he is a retired lobster fisher. Thinking himself so much smarter than a lobsterman, the lawyer offers to play a game with him. "Here's the deal, Earl," says the lawyer, "I'll ask you a question. If you can't answer it, you give me five dollars. Then you ask me a question. If I can't answer it, I give you five hundred dollars." Earl agrees to the game. The lawyer's question to him is: How many kilometres is it from earth to the moon? The fisher doesn't know so he pulls out five dollars and gives it to the lawyer. Now it's Earl's turn. His question for the lawyer is: How long can a lobster live if it never gets trapped? "Hmmm…oh my, you got me there, Earl!" says the lawyer, as he reluctantly hands over five hundred dollars. "So tell me Earl," he asks, "how long *can* a lobster live?" The ol' fisher reaches in his pocket and hands the lawyer another five dollars.

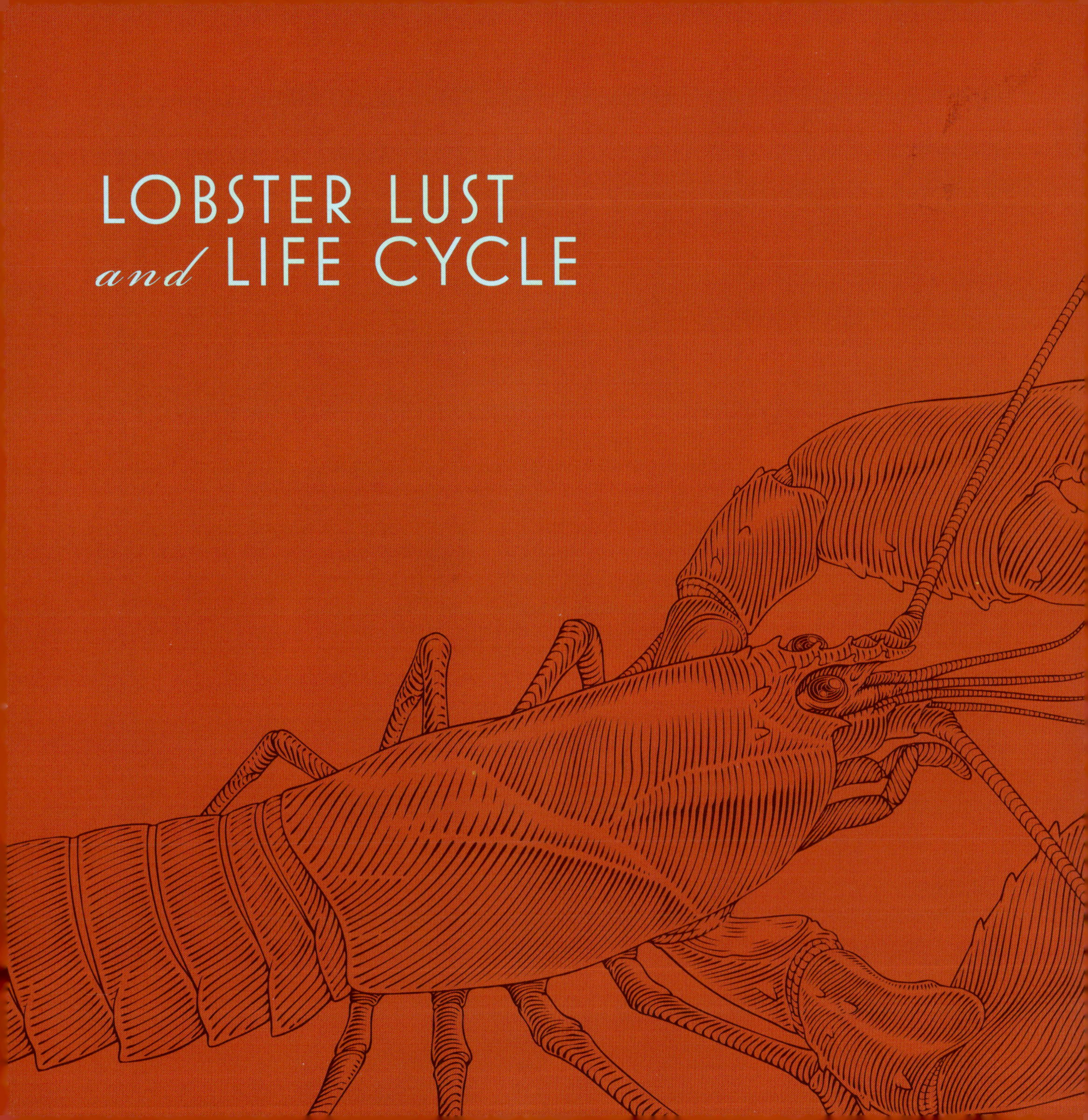
LOBSTER LUST
and LIFE CYCLE

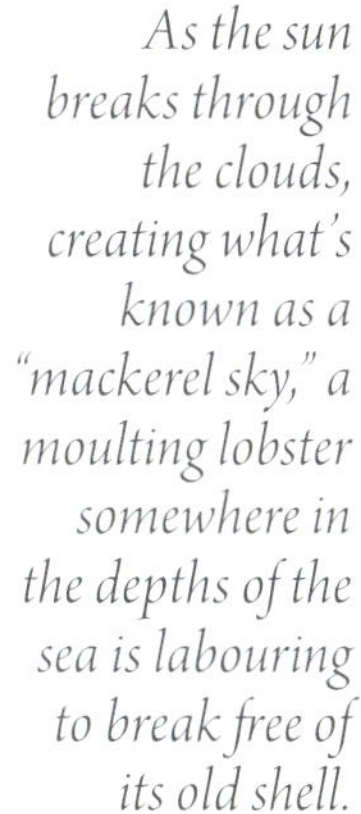

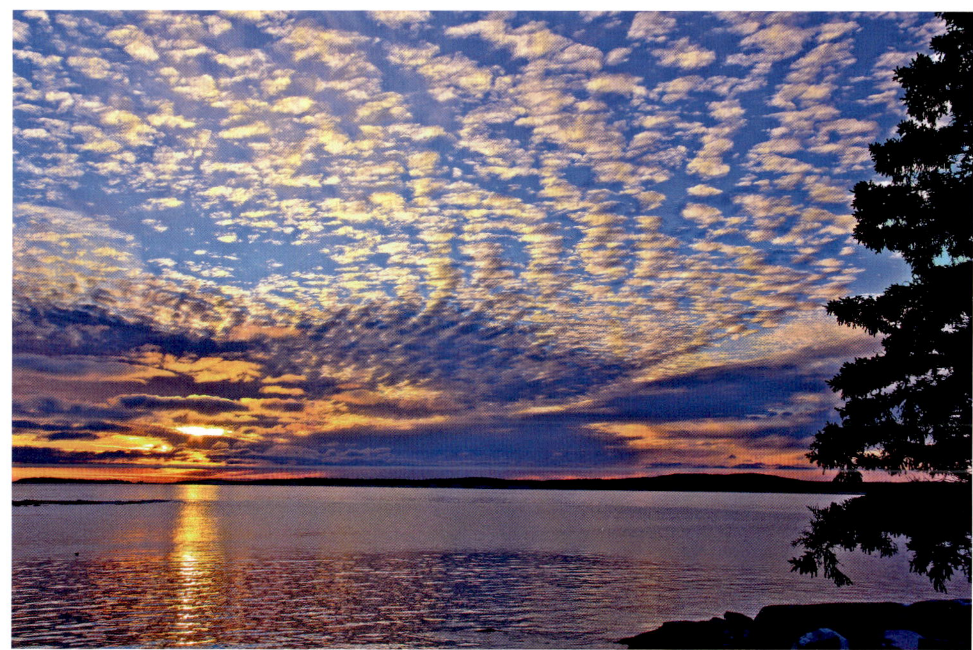

As the sun breaks through the clouds, creating what's known as a "mackerel sky," a moulting lobster somewhere in the depths of the sea is labouring to break free of its old shell.

Bursting At The Seams: Moulting

Imagine a snug-fitting pair of jeans. As middle age sets in, there comes a time when those favourite jeans just get too tight and must be replaced with the next size up. That is somewhat what happens with lobster growth. A lobster is always growing, but its shell remains the same size until something has to give. The outer layer just gets too uncomfortable and must be shed. When a lobster has outgrown its old shell (the exoskeleton), it must wiggle its way out to allow its soft outer body to expand and harden into a new, larger shell. Hormones secreted from glands inside the lobster's head urge it to exit its old shell. The lobster then seeks a safe hiding place, as it is most vulnerable during and shortly after a moult.

If all goes well during the moult, the lobster will free itself of its old carapace within twenty to thirty minutes. Given the hard, sharp spikes and shell edges

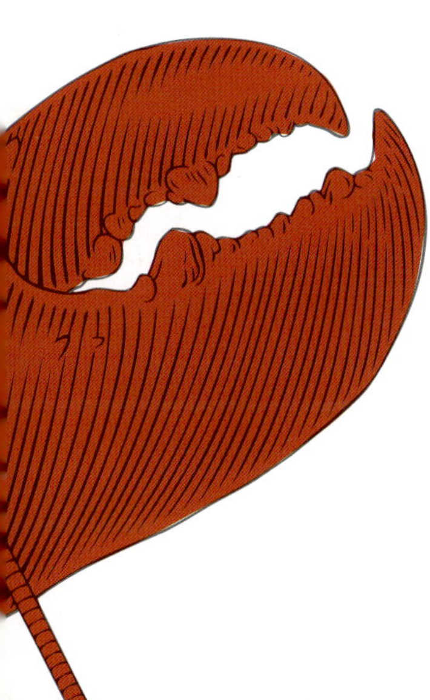

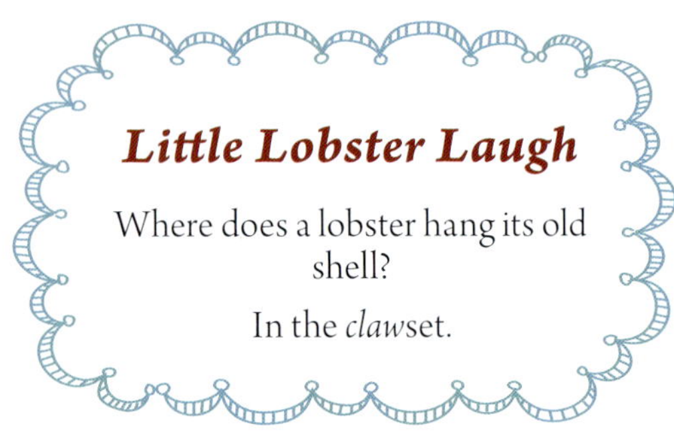

Little Lobster Laugh

Where does a lobster hang its old shell?

In the *claw*set.

involved, this is a very risky operation. The lobster's delicate young shell layer could easily get punctured or torn on the way out. It's not uncommon for lobster to suffer severe injuries, deformities, or even die as a result of a moult gone wrong. How those large claws get through the much smaller narrow knuckles is somewhat baffling.

Here is the breakdown on how a lobster moults: In the weeks prior, the lobster begins to grow a new skin layer under its shell. Hormones cause the lobster to pump seawater in between its flesh and the hard old shell using what is called "hydrostatic pressure." This is the pressure exerted by fluid in the confined space between the lobster's "new skin" and the old shell. They begin to separate. The lobster then lies down on its side and begins to go into "labour." After lots of quivering and body twists, a break first occurs where the thorax (upper body) meets the tail (abdomen). The tail's shell is the most straightforward part of the exoskeleton to come off. The lobster's greatest challenge is shedding the upper-body shell, with its eight little side legs,

Frequency of Lobster Moult

How often a lobster moults depends on its age. Young lobster shed their shells more often than older lobster. Before a lobster reaches one year of age, it will have moulted thirty-five times in the egg, four times in the water column at the planktonic stage, once more when it settles as a minuscule lobster on the sea floor, and continues to moult about once a month for four or five months. This comes to a total of about forty-four moults in that first year of life. In the second year, a lobster will moult every other month four more times, at which point the time between moults increases. At age three and four, the lobster will typically moult two to three times each year. Five- and six-year-old lobsters moult once or twice yearly. By age seven, the lobster will moult but once yearly. From then on, it may moult only once every two or three years.

Autotomy

Autotomy is the ability in some animals to self-amputate an appendage, usually a leg, tail, or antenna, in order to escape the grasp of a predator. It is most common among amphibians, reptiles, and insects. In lobster, self-amputating a claw or a leg is usually a response to stress: water that is too close to freezing or too warm, fights between rival males, and moults gone wrong are the most common causes of autotomy. Regeneration of the appendage begins soon after, but for lobster it is very slow. As an example, if a three-year-old lobster looses a claw, it will take at least three more years for the new claw to grow proportionate to the other one.

two large claws, head, and thorax. This is where many things can go wrong.

The claws are especially difficult to wiggle out of the old shell. Getting a claw stuck in a knuckle shell is the most common injury for lobster when moulting, and can lead to fatal infection. When this happens, the lobster will usually opt to abort the limb (autotomy) and grow a new one. If both claws are amputated, there is a good chance it could lead to starvation, as lobster need their claws to break up food and bring it to their mouthparts. A clawless lobster will have to search for soft food sources until it grows a new set, and claw regeneration is very slow.

After much struggling to wiggle every body part out of its old shell, the lobster finally breaks free. The entire upper body, side legs, claws and all, emerge like a closed jackknife. The lobster's body experiences a short-term loss of integrity. It then slowly stretches out its limbs and, soon after, rights itself. It will remain in hiding for little more than a week to allow the new shell to expand and harden. Lobster are most susceptible to predation during this period, and will only come out of hiding if they smell food nearby. Even bait. Yes, a hungry soft-shell lobster will take its chances and venture into a trap for an easy meal.

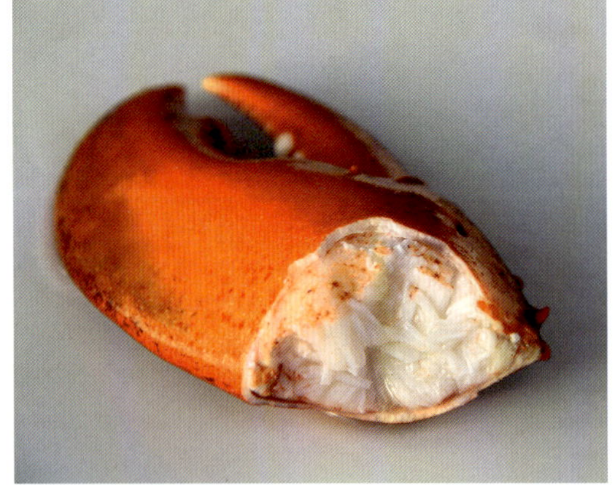

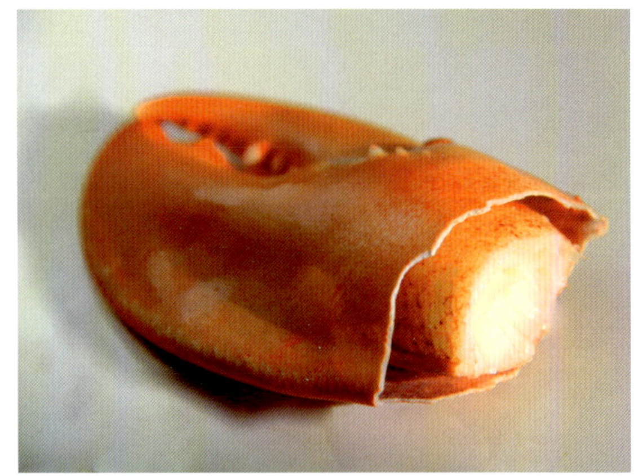

These cooked claws clearly show the difference in quality between hard-shell (left) and soft shell-lobster (right).

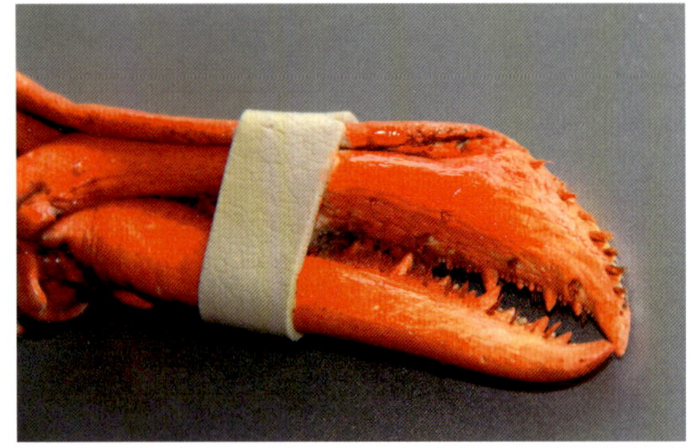

This claw fold is the result of a lobster being banded while still at the soft-shell stage following a moult.

Soft-shell lobster are not as valuable as their hard-shell counterparts. The meat inside soft shells is compromised because it is so full of water; it shrinks when cooked and the texture tends to be rubbery. Additionally, soft-shell lobster are not suitable for shipping because they are prone to injury from general handling. Their claws get deformed when banded. Hard-shell lobster may be more work on your plate, but the taste is well worth the extra effort.

A lobster's growth is most rapid in the first couple of weeks after a moult. It can put on more than a 110 grams (1/4 pound) per moult. Lobster in captivity have been observed eating their old shells; it is thought they do so to replenish lost calcium in order to harden the new shell faster.

She-Lobster Seeks Fine Male Specimen: Procreation

Have you ever wondered *how* they do it with that heavy-duty clunky outer armour? Those oversized boxing glove–like claws, spiky horns all over, and sharp double-edged tail? It's like the age-old question: how do porcupines manage to copulate without stabbing each other? The answer is, well, *very, very carefully*. The same can be said about lobster, but it's a different mechanism altogether than we mammals. The way lobster procreate is an ingenious marvel of nature that has only recently come to light.

Evolution has devised the perfect reproductive system for what appears, on the surface, to be impossible. Oceanographers observed lobster courtship rituals by placing males and females in large aquariums stocked with many of the natural features of their natural ocean environment. Here's the juicy part: The act is reported to be remarkably human-like. There is

Farm-Raised Lobster Research

Lobster data collection didn't begin until 1966, when American demand for lobster was high and supply was dwindling. This gave rise to the US Sea Grant Program, which funded lavish research projects aimed at better understanding the life cycle and reproduction habits of lobster. The main motive was to "domesticate" the wild American lobster in the hope of farming its progeny as a commodity. In 1974 a similar Canadian initiative was conducted in the coastal town of St. Andrews, New Brunswick. A large study on raising lobster brood stock in artificial settings was short lived. It was concluded that the experimental lobsters were not only inferior to those in the wild, but also fiscally unsustainable. Each larvae stage requires different conditions. Many specialized holding tanks are needed. The main obstacle is that young lobsters are highly cannibalistic and so must be kept separated from one another while growing. It takes immense amounts of space, water, and filtration technology and it takes seven years for a lobster to reach minimum market size. Together, these factors make farm-raised American lobster unsustainable.

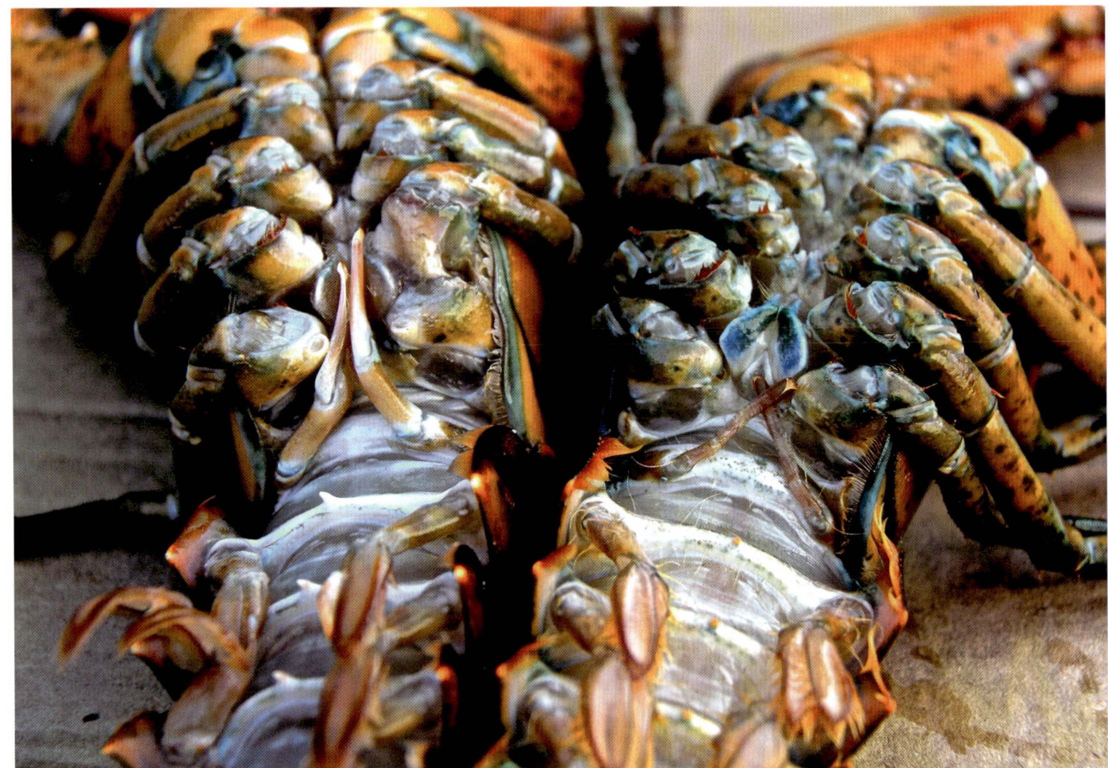

Male swimmerets (left) and female swimmerets (right)

a cheek-to-cheek embrace. They spend more time getting "in the mood" than "in the act"—which takes no more than ten seconds! Keep in mind, though, that they have no genitalia (as we know it).

To begin, there are three ways to tell a female lobster (hen) apart from a male lobster (cock). Side by side, when the same length, a female's tail (abdomen) is wider than a male's; this allows her to carry thousands of eggs. Also, on the underside of a female's tail are barely visible hairs that help secure her minuscule eggs. Weight for weight, the male has larger claws than a female. The most common method for telling male and female lobsters apart, however, is to look at the under-tail. On a male, the first two little legs of the five pairs, called

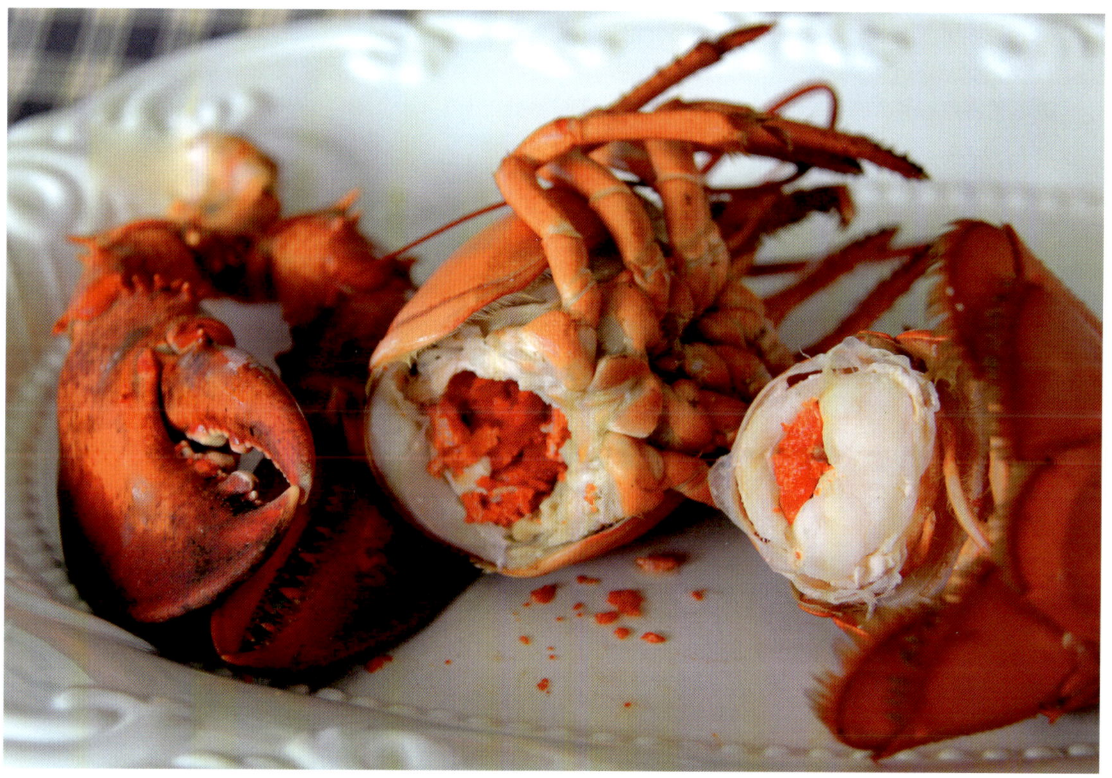

Many lobster enthusiasts consider red lobster roe (unfertilized eggs) a delicacy.

swimmerets, are sharp and stiff. On a female, the first two swimmerets are flexible and feathery.

That flavourful, brilliant red aggregation that extends from the head to tail of a cooked female lobster is a contained bundle of unfertilized eggs—the equivalent of a human woman's ovaries—called "roe." Fertilization can only take place when these eggs come in contact with a male lobster's spermatozoa.

The mating rituals of lobster are fascinating. When a sexually mature female senses she has outgrown her shell, she goes into heat. As the one who must seek out a suitable mate, the female lobster releases powerful sex hormones, called pheromones, into the surrounding

Lobster PMS

In the late 1990s American Researcher Diane Cowen observed that just before moulting for copulation, some female lobsters were easily agitated—moody, you might say. They displayed visible aggression toward other nearby female lobsters. Cowen coined this condition PMS: "pre-moult syndrome." According to Cowan, "before they moult [females] have an activity peak and can go a little berserk." She says that other hens may be the target of attacks just before a female lobster goes into moult.

water. The male has to find ways of attracting a female to him—and away from rival males in the area. Male lobster competition can result in injurious fights: a crushed claw, missing antennae, cut side legs, even a severed eye.

Despite their miniscule brains, lobster have mating rituals as elaborate as birds of paradise. To attract a female, the male lobster needs to find the best "love den" possible, usually between or underneath large rocks. In order to lure a female in, he places relics of his best catches—items like mussel shells, hollow crab bodies, and empty sea urchins—at the entrance of the "love den" to show potential mates he is a fit and competent hunter of desirable stock.

When a female lobster finally settles on a male and his enticing abode, she enters the dimly lit room. A sex hormone triggers her to moult and she begins to shed her shell in the protection of the den. While she undergoes a twenty- to thirty-minute "strip-tease," the male guards the den's entrance. A lot of biochemistry flows through the water between the male and female lobster at this time. This is when their antennae come in handy.

Lobster have three pairs of antennae: a set of two long antennae and two pairs of smaller antennae called antennules. Since lobsters live in low-light environments and are most active

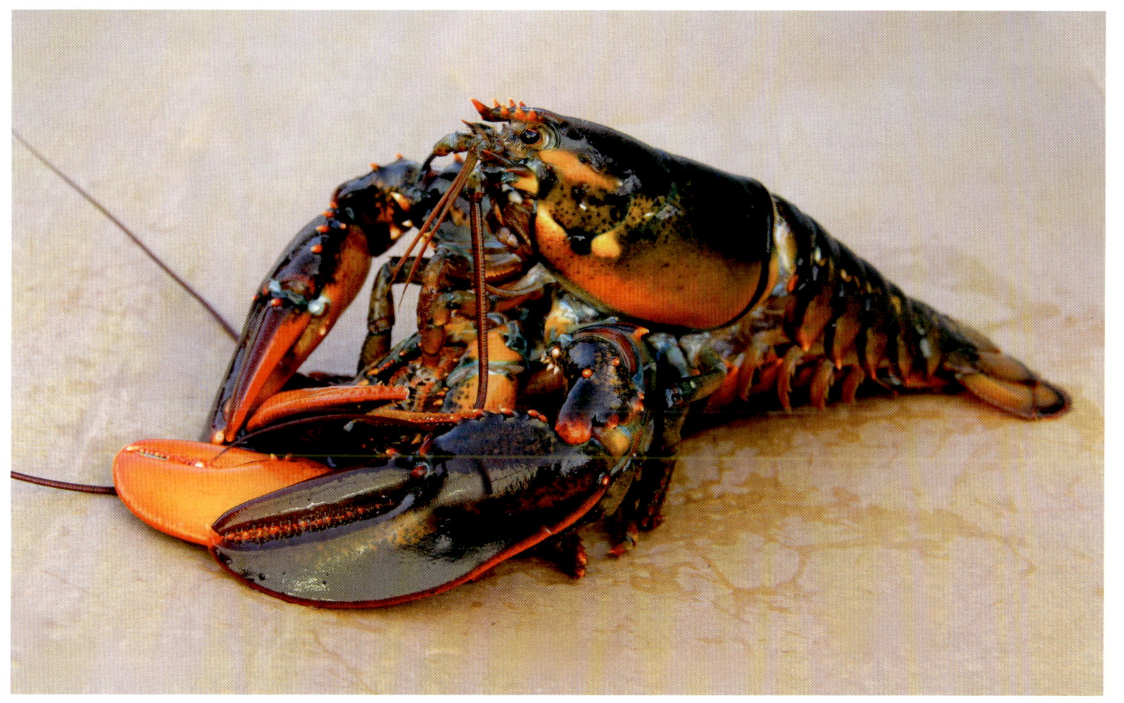

How a male lobster positions himself over a female to transfer his DNA to her.

at night, they feel their way around using their long antennae. They have very poor eyesight, so the long antennae are important spatial sensory organs. The antennules are scent receptors; lobster flick them up and down vigorously to "sniff" and "smell" their environment. This is how they find food. But the antennules are especially important in detecting chemical signals emitted by a female. The knight in shining armour knows, from the scents released by the lovely soft-bodied princess, when she is ready to receive him.

As soon as the male gets the "let's-go-for-it" hormone signal, he re-enters the den and uses his long antennae to determine his mate's position. Once he locates her, he appears to sooth and reassure her with gentle taps of his antennae on her tender new skin. The male mounts the soft-bodied female ever so carefully. Then, using his side legs, he very delicately flips her over onto her back.

Lobster Life Cycle

Only 1–10 lobsters are expected to survive to market size (1–3 pounds) out of about 50,000 egg hatchlings.

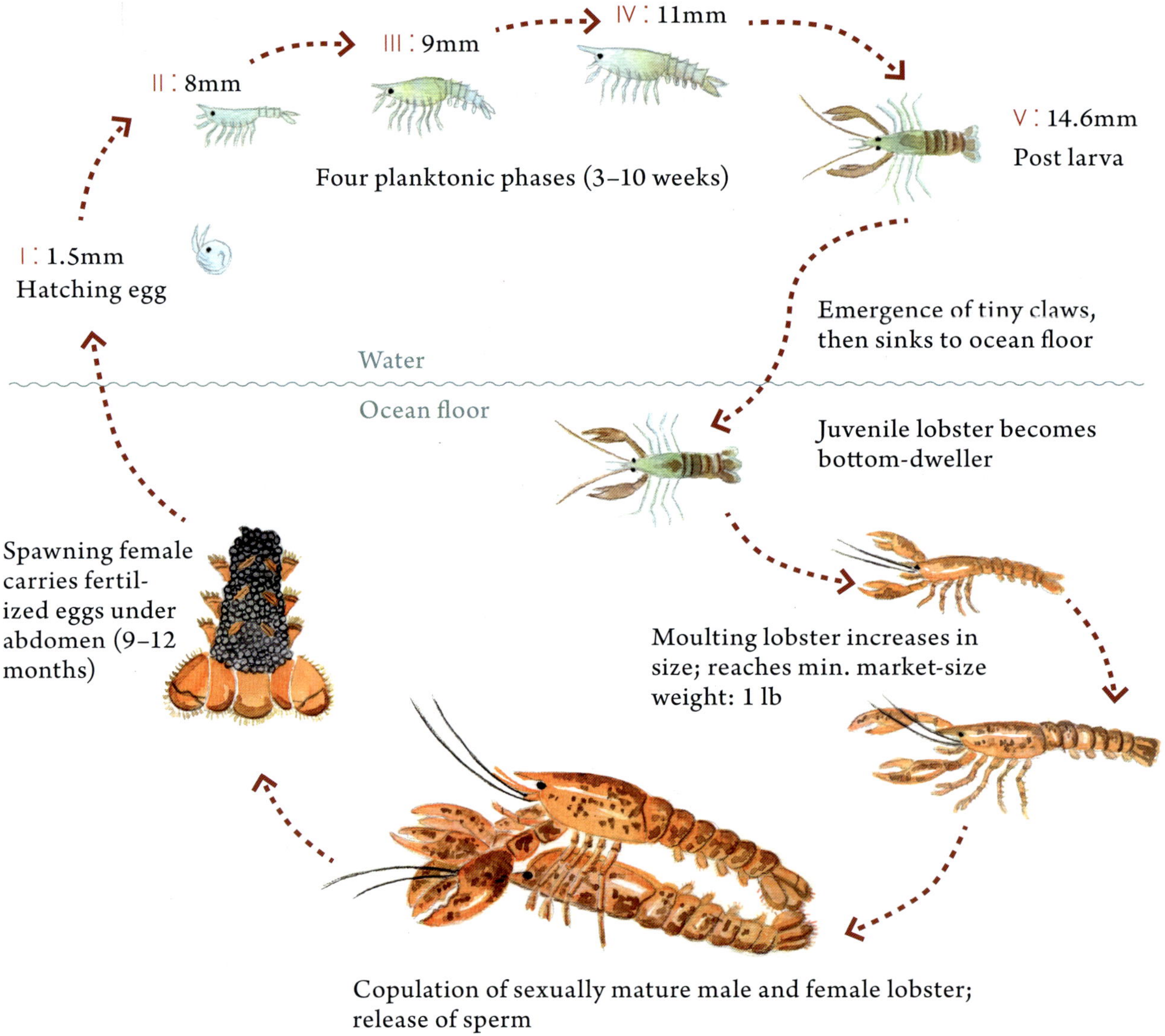

I : 1.5mm
Hatching egg

II : 8mm

III : 9mm

IV : 11mm

Four planktonic phases (3–10 weeks)

V : 14.6mm
Post larva

Emergence of tiny claws, then sinks to ocean floor

Water

Ocean floor

Juvenile lobster becomes bottom-dweller

Moulting lobster increases in size; reaches min. market-size weight: 1 lb

Copulation of sexually mature male and female lobster; release of sperm

Spawning female carries fertilized eggs under abdomen (9–12 months)

Just below the female lobster's swimmerets is a little pouch on the exterior of her abdomen called a seminal receptacle, where she will store the male's genetic material. The male's job is to somehow transfer his sperm into it. Male lobsters do not possess a "reproductive organ." The male lobster uses his specialized rigid swimmerets to pry the female's receptacle open. Then, in a lusty face-to-face grasp, he squirts his DNA, in the form of microscopic sperm bundles, into the female's pouch.

Timing is everything in the sex lives of lobster. If mating takes place before the female's moult is complete, fertilization will be unsuccessful, as the sperm will go with her old shell. If

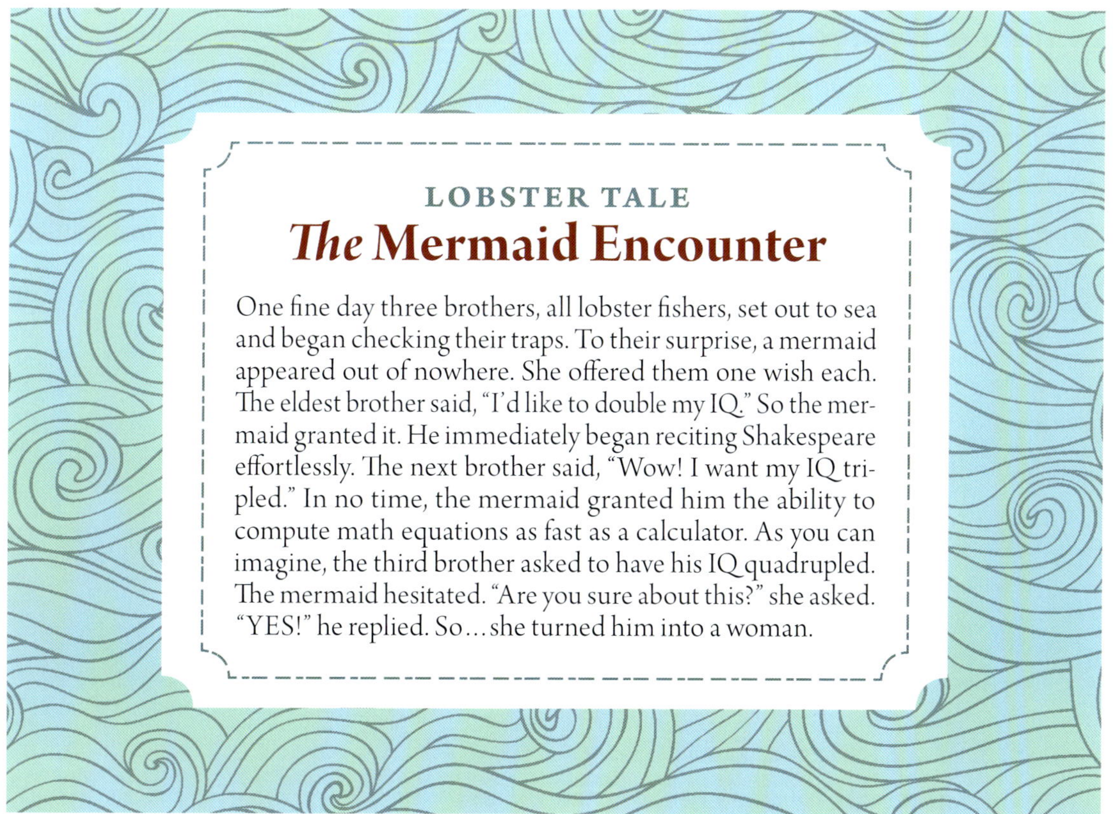

LOBSTER TALE

The Mermaid Encounter

One fine day three brothers, all lobster fishers, set out to sea and began checking their traps. To their surprise, a mermaid appeared out of nowhere. She offered them one wish each. The eldest brother said, "I'd like to double my IQ." So the mermaid granted it. He immediately began reciting Shakespeare effortlessly. The next brother said, "Wow! I want my IQ tripled." In no time, the mermaid granted him the ability to compute math equations as fast as a calculator. As you can imagine, the third brother asked to have his IQ quadrupled. The mermaid hesitated. "Are you sure about this?" she asked. "YES!" he replied. So…she turned him into a woman.

a soft-shelled male attempts to mate, he will be unable to open the female's pouch; his swimmerets must be stiff and sharp to get the job done.

When the task is complete, the male lets go and the female's pouch self-seals. She now has her very own little "sperm bank." Mission accomplished, the female lobster will remain in the crevice for about a week, until her new exoskeleton has sufficiently hardened. Then she will go on her way and the male lobster will once again prep his "love den" in the hope of attracting another mate.

Underage Drifters: Spawning

Although she has mated, the female lobster is not technically pregnant until her eggs have been fertilized. Once her new shell has hardened, it is up to her to choose when and where to spawn. Much mystery surrounds lobster spawning. Divers have reported sighting large gatherings of female lobsters spawning together in calm, shallow inlets. It appears that late summer is the preferred time to lay their eggs.

Lobster Sabotage

There is an unfortunate and very illegal practice among greedy fishers called "scrubbing," whereby eggs are scraped off a brooding lobster with a brush so she can be sold. The common belief is that these fishers are "cutting their own throats": by participating in this practice, thousands of potential lobster are wiped out from the future of the fishery. A more common but less grievous practice among harvesters is to conceal the berried female in the rest of the catch so they get paid for their trouble. It is then up to—and in the best interest of—lobster pound workers to release her back to sea, in tact with her brood, to keep the species going.

When a tranquil, safe hideaway is found, a lobster hen turns over on her back to begin spawning. She curls her tail up to form a sort of ladle. A hormone signal causes her sperm-filled seminal receptacle to open, and she gradually releases all of her eggs over the pouch opening. The female lobster fans the preserved sperm over the eggs with her feathery swimmerets and guides the now-fertilized eggs into her curled undertail. A glue-like excretion from the swimmerets firmly attaches the fertilized eggs to her abdomen, where they will remain safe from predation as they develop.

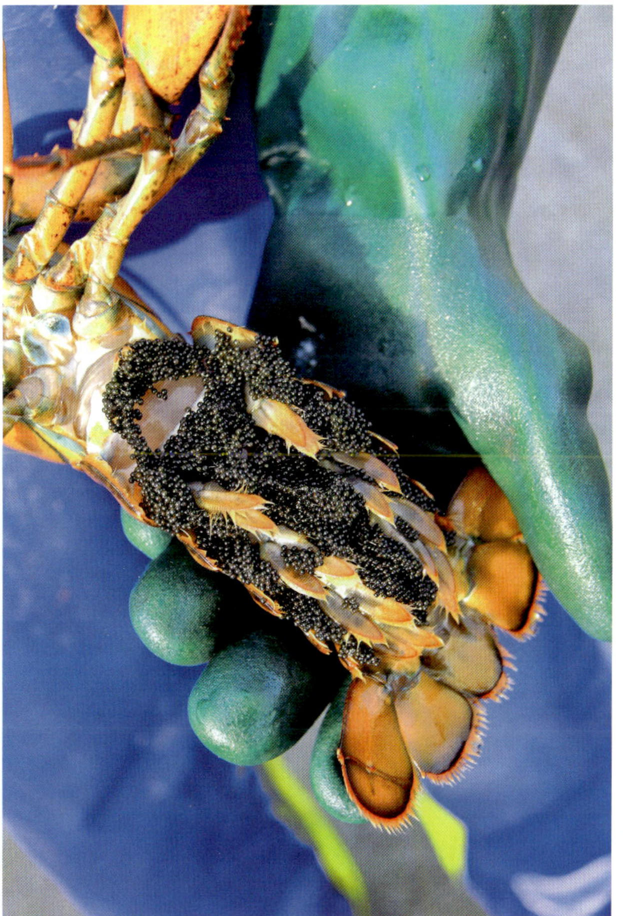

Brooding female lobsters are called "berried" because the egg mass resembles raspberries or blackberries.

An egg-laden female is called a "berried" or "brooding" lobster. In both Canada and the US, it is strictly prohibited to keep a trapped "berried" female. She must be promptly returned to sea unharmed. This is the future generation after all. Returning large females and egg-bearing females to sea is key to a sustainable lobster fishery. The protection of berried lobsters, which dates back to 1870, was the very first conservation measure put in place in Canada.

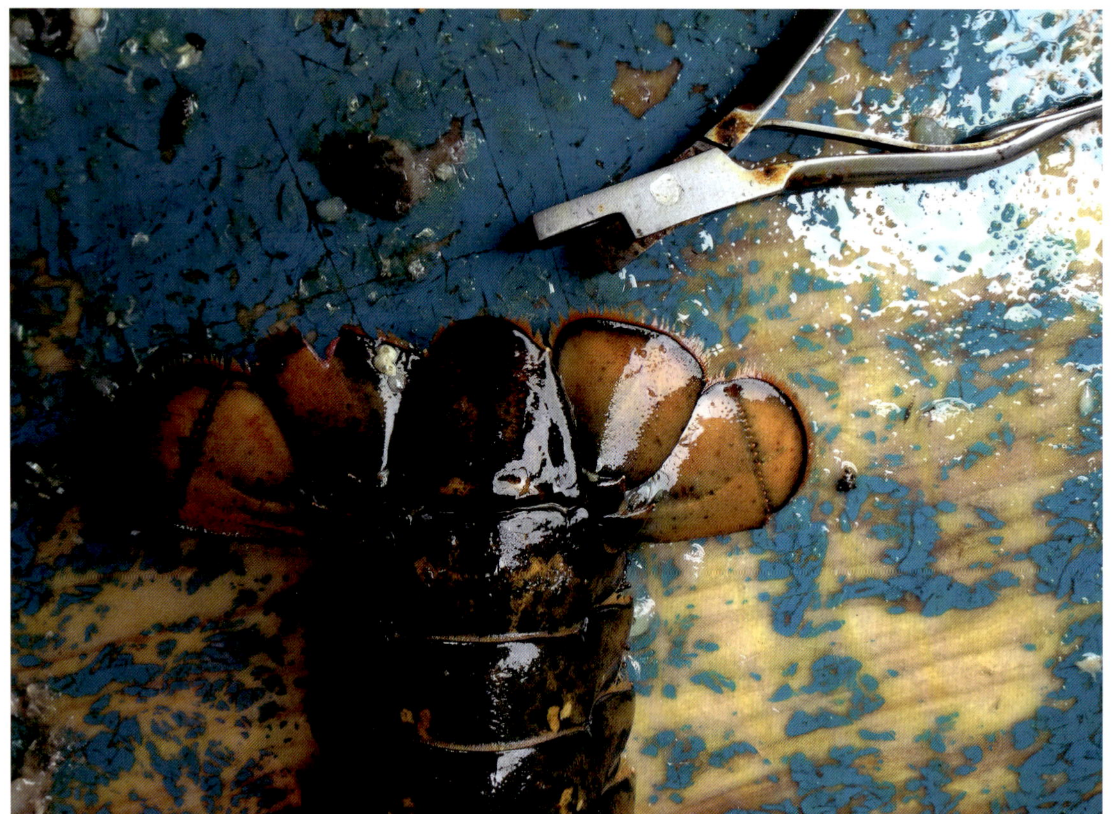

A notched productive female lobster and notching tool.

Only between one and ten lobster hatchlings will make it to adulthood. Fortunately, lobster are very prolific breeders. Young females (20cm/8") produce approximately 5,000 eggs per spawn. Mature, larger females (50cm/20") produce ten times that amount of eggs—about 50,000 and more each year as they age! That said, one large female lobster produces as many eggs as ten young female lobster. For this reason, well-informed, conscientious lobster fishers usually return large females (2 1/2 pounds or more) to sea whether they are berried or not, because these gals will produce copious amounts of juveniles and in return keep lobster stocks strong.

A recent conservation practice, which began in the United States, is to "notch" the tail tip of female lobster that are returned to sea. This indicates to other harvesters that she was caught before and, if caught again, she is to be thrown back to ensure the future of the population. A notched female should not end up on your plate. This has proven difficult to enforce, however; in Canada, notching is voluntary.

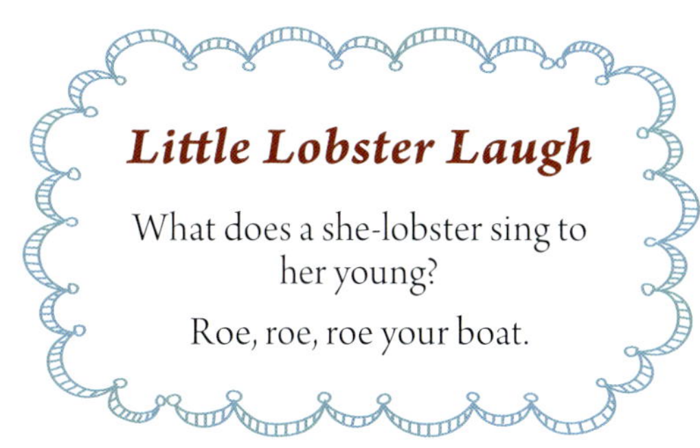

Depending on water temperature, a brooding lobster will carry her eggs for between nine and sixteen months, with accelerated egg development in warmer, shallower waters. One by one, the eggs break open as they come to term. Thousands of lobster larvae rise to the surface of the sea to become part of the flow of plankton. Lobster larvae are zooplankton (animal-based organisms) that feed on phytoplankton (plant-based organisms).

Planktonic lobster larvae are called "drifters." They do not resemble their clawed parents; they look more like tadpoles with large dark eyes. Few will survive predation. Many marine creatures feed on lobster larvae, such as barnacles, jellyfish, sea sponges, small fish, shrimp, krill, and even whales. Baleen whales consume copious amounts of plankton. They swim with their mouths wide open to collect food, and then filter the water out through their baleen (feathery mouthparts) to separate it from their food source.

Planktonic lobsters that survive predation will metamorphose several times, getting a bit bigger as they go, for about two weeks. It is not until the post-larval stage that they take on the morphology of a true, miniscule lobster with claws (about 1/4 inch), at which point they sink to the sea floor. This is when they begin a whole new life as bottom-dwellers like their parents: searching, scavenging, and hunting for food. Juvenile lobsters nibble on sea plants and feed on tiny aquatic animals living on and among those plants: worms, mollusks, sea snails, and fish eggs, for example. These young lobsters also feed on trap bait but fortunately are small enough to escape through the mandatory exit vents or from between the laths.

Ode to Phytoplankton

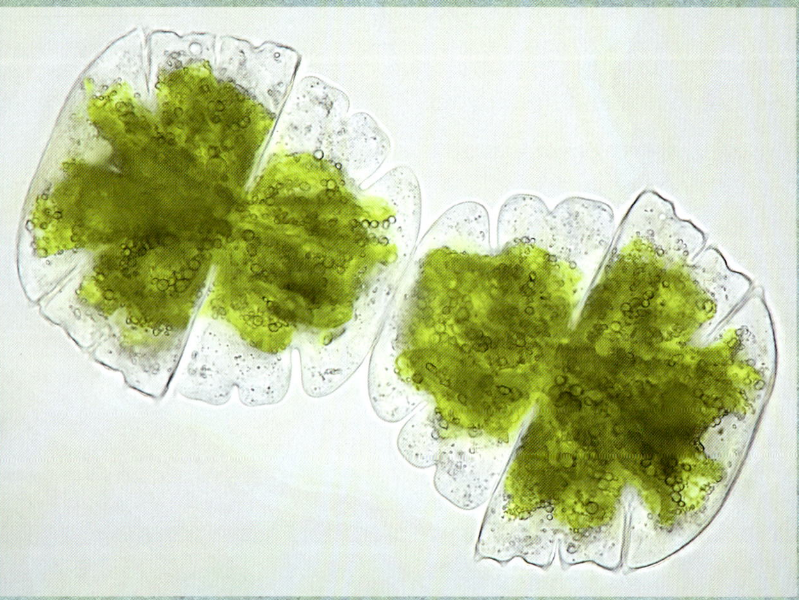

Algae, a common type of phytoplankton.

Phytoplankton are the first plant-based organisms to have been born out of the primordial soup that once covered our very young planet. They are the ancestors of all plant life, from algae, mosses, and grasses, to flowering plants and trees, including the vast diversity of aquatic plants that colonize the oceans. They are invisible to the naked eye. They are food to organisms too small to see. Yet, phytoplankton are the very the foundation of the food chain.

These silent and unseen marine organisms produce nearly half of the air we breathe; the rest comes from land plants. What forests and phytoplankton have in common is that they both derive most of their energy from sunlight, by a process called photosynthesis. They both absorb carbon dioxide and release life-sustaining oxygen in return.

Unlike terrestrial and marine plants, phytoplankton are mobile. The name is derived from the Greek word *planctos*, meaning "wandering" or "drifting." Interestingly, phytoplankton have no swimming abilities; they have no need for fins, gills, or tails. Instead, they depend entirely on currents and tides for transportation. They travel immense distances throughout the oceans of both hemispheres.

Depending on the time of year and geographical area, a teaspoon of seawater can contain as many as two thousand species of phytoplankton blooms. When conditions are favorable, phytoplankton blooms grow in such concentrations that they are visible via satellite as fluorescent green swirls over the ocean surface. Other than this, the only way to see phytoplankton is under a microscope.

These single-celled plants are extremely diverse. Most are intricate, beautiful, symmetrical designs. Others are simple ribbon-like shapes. They live only a few days but their impact on the diversity of life on earth cannot be underestimated. These microscopic marine plants are the umbilical cord to all other life forms, including humans. Without phytoplankton, life as we know it could not exist.

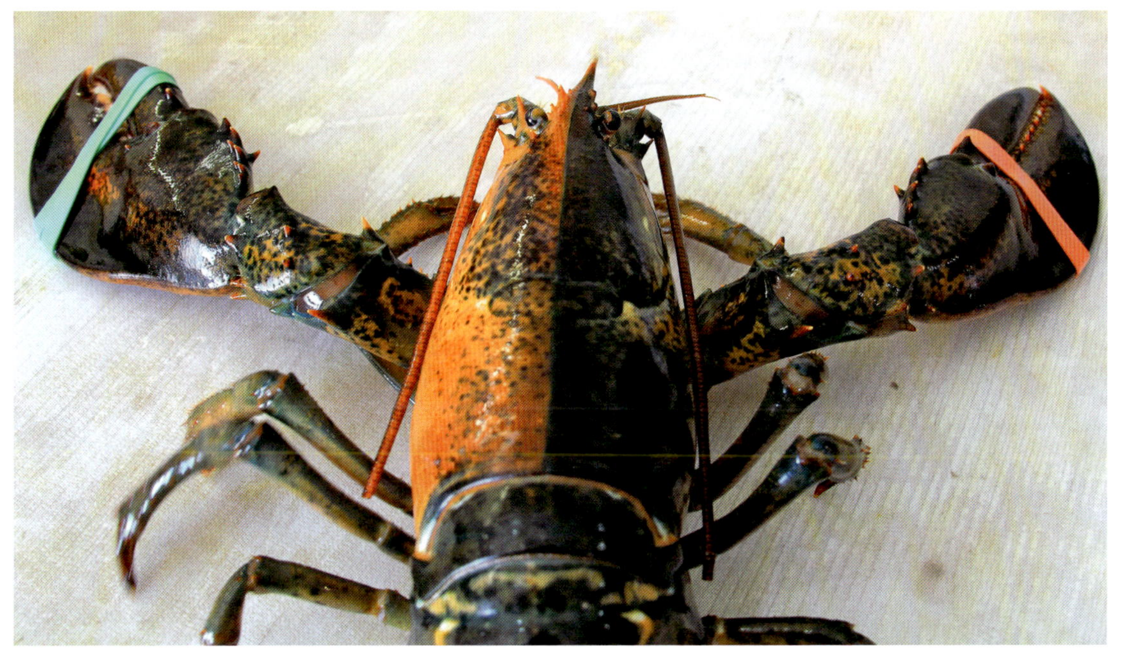

This rare two-toned lobster showed up at Ryers lobster pound in Indian Harbour, Nova Scotia, in the spring of 2015.

When undersized lobster are found in a trap, harvesters must throw them back to sea or risk facing stiff fines (two thousand dollars per infraction).

Lobster Oddities

Now and then, a one-in-a-million lobster with bizarre colouration shows up in a trap. Blue lobsters, in particular, get a lot of attention. Those who've seen one say it is the prettiest blue they've ever seen. Unusual pigmentation is a result of a genetic mutation that causes an excessive amount of a particular protein (or lack thereof), which affects shell colour.

No matter the anomaly, all lobsters turn red when cooked, except albino lobsters, which remain white as they lack pigmentation. Lobster meat flavour is not affected by shell

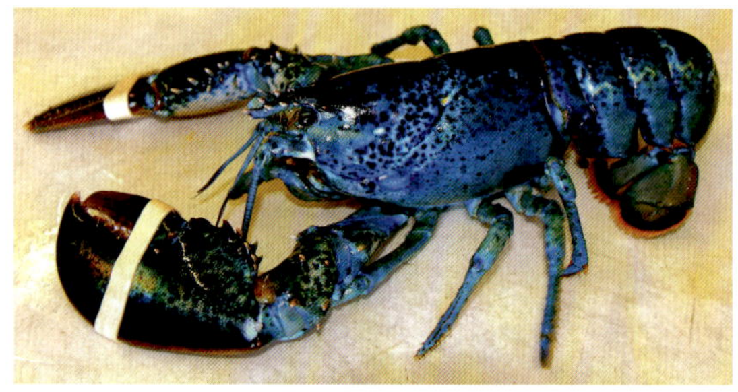

This Nova Scotia–caught electric-blue lobster gets "rock star" attention at Ryers Lobster Retail.

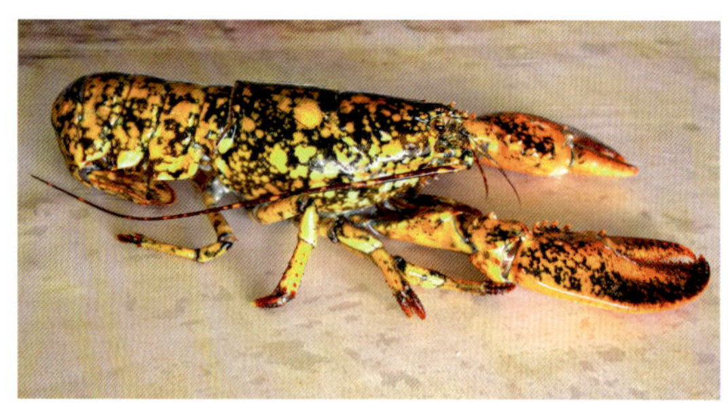

Black and yellow-orange mottled lobsters are called "calico."

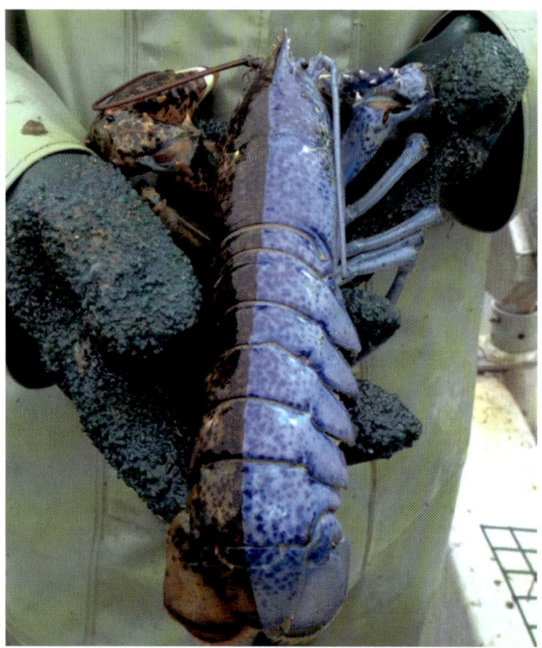

On June 17, 2013, Peter Marche caught a never-before-seen half-blue lobster near St. Georges Bay, Newfoundland..

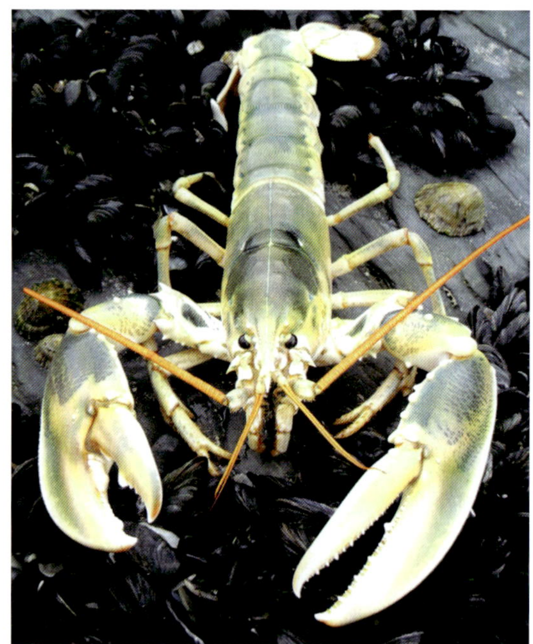

This so-called "ghost lobster" is an albino. It lacks the normal pigmentation of other lobsters so it remains whitish in colour even when cooked.

colour—even so, novelty lobsters rarely end up on someone's plate. They tend to be showcased live with the intent to promote interest and sales. Odd lobsters are usually released back to sea at the end of the tourist season, or as soon as they appear to show signs of stress, such as listlessness, leaning to one side, foaming at the mouth, or self-amputation.

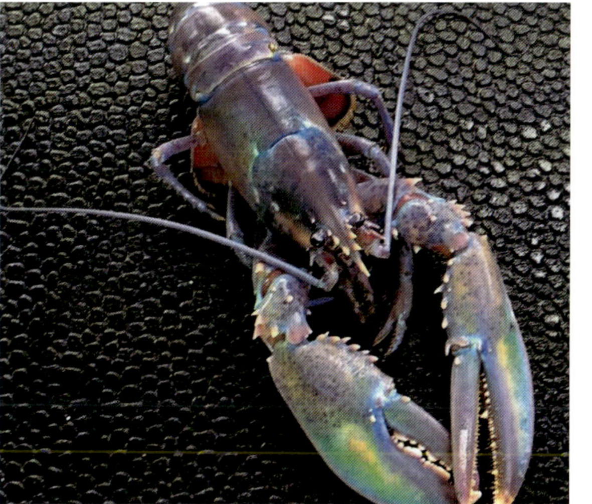

A quick-thinking lobster harvester took a photo of this unique lobster colouration. The "rainbow lobster" was caught and released near Brier Island, Nova Scotia, in December 2015.

Another shell-colouration anomaly is what fishers call a "two-toned lobster." These lobsters have a clearly defined split: red on one side and traditional greenish-black on the other, as though the lobster is sunburned on one side. The split can go from "nose" to tail tip. Even rarer are lobsters split in four, with the head and body split red/green and the tail split green/red. The odds of discovering one of these in a trap are 1:100,000.

Recently, a full-spectrum coloured lobster was found off the coast of Nova Scotia. It's been called the "rainbow lobster," the "psychedelic lobster," and the "technicolour lobster." Captain Chad Graham caught the unique crustacean on December 19, 2015, from the waters of St. Marys Bay, near Brier Island, Nova Scotia, where he keeps his boat, the *Chad & Sisters Two*. He took a photo of the odd-looking crustacean, and then promptly returned it to sea because it was undersized. It wasn't until a few days later, when he downloaded the picture from his phone, that he realized what a spectacular specimen it was. There is no other record of such a lobster colouration anomaly. The photo has received a lot of attention on social media; it even reached a biologist who contacted Mr. Graham in hope to conduct research on the lobster in order to better understand its bizarre pigmentation. Unless the little "rainbow lobster" gets caught again at legal size, we may never learn exactly why it displays such an array of iridescent colours.

And then there are lobsters that exhibit hermaphroditism: half male and half female. This condition appears to be on the rise in lobster and other aquatic and amphibian creatures and its cause is not well understood. Although it is very, very rare in all living things, in humans hermaphroditism is the result of either a genetic mutation or abnormal embryo development. When this condition occurs in frogs, it is suspected to be the result of environmental factors such as a sudden change in temperature, pollution, or a poor diet in egg-bearing females. Since an egg-bearing lobster has never been found with both male and female swimmerets, we can only assume that reproduction is impossible for hermaphroditic lobsters.

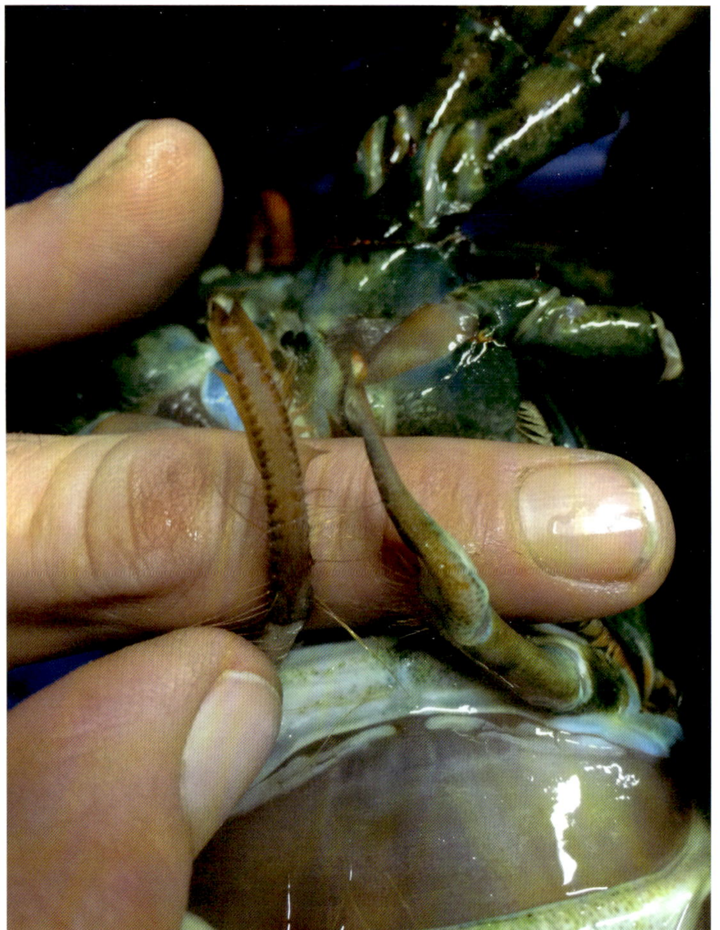

Lobsterman Ken Cleveland took this photo of a lobster with a female swimmeret on the left and a male swimmeret on the right.

LOBSTER TALE

Adventure *of a* Tourist

A tourist loved his meal of Atlantic-caught lobster so much, he bought fresh lobsters to take back home for his family. He was handed a box of live lobsters carefully packed in ice. Once he arrived home, he opened the box and his jaw dropped. He was dismayed to find dark-greenish–looking lobsters. He assumed they had spoiled during the trip and discarded them!

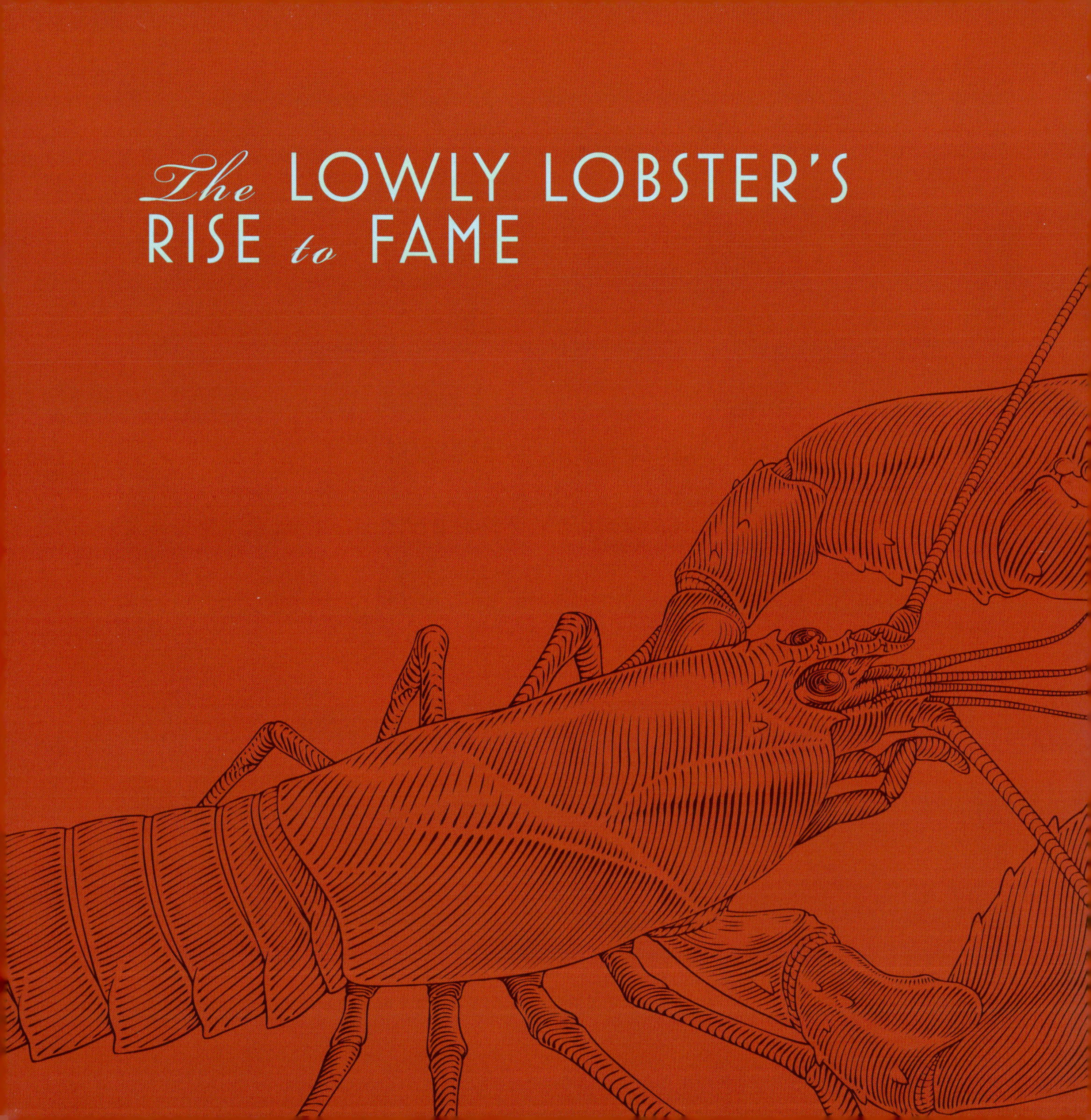
The LOWLY LOBSTER'S
RISE to FAME

In a very short period of time, lobster as we know it today rose to prominence—above the almighty cod—as one of the world's most sought-after seafood delicacies. The appetite for cold water–caught Atlantic lobster has grown exponentially in just the last few decades. The demand for live, rather than canned, lobster resulted in lobster's greatest ever increase in value. Lobster-stocked aquarium displays, the hallmark of freshness from the sea, are a relatively new marketing strategy. Not only is the lobster fishery an export industry, it remains a major tourist draw in Atlantic Canada. Lobster's unique, briny-sweet taste has become highly desirable at home and abroad.

Reign of the Lobster; Demise of the Cod

The rise of lobster to such prominence is really a rags-to-riches story. In fact, it is well documented that lobster was long considered a food for the poor. In her book *How to Catch a Lobster in Down East Maine,* author Christine Lemieux describes the perception of lobster as a subsistence food at the turn of the twentieth century. Not only that, lobster were so commonplace and plentiful to coastal communities they were raked up at low tide along with seaweed to be mixed into the ground as fertilizer.

The seafood of choice prior to the rise of lobster was codfish. Lobster and cod had very similar beginnings. Both are reputed to have been extremely abundant in the North Atlantic waters of the New World. At the point of European contact (mid-1400s), both lobster and cod were colossal in size. It wasn't uncommon to find lobster measuring 1.5 metres in length and weighing as much 55 pounds. Cod harvested in the mid-1600s were recorded to have weighed about 220 pounds each. Today, codfish grow to a meager 11 pounds at most, while the average lobster harvested today is just under 1–2 pounds.

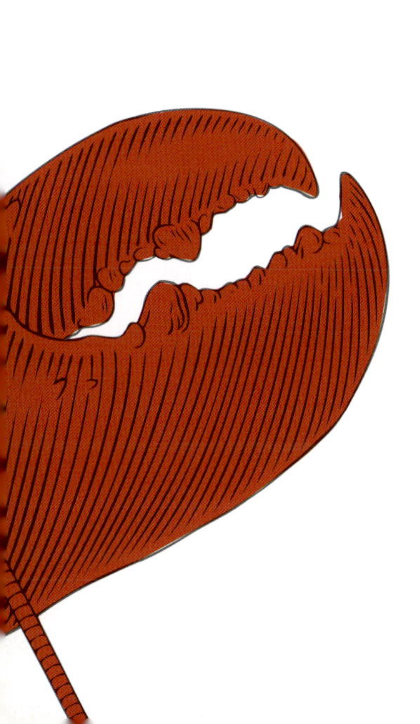

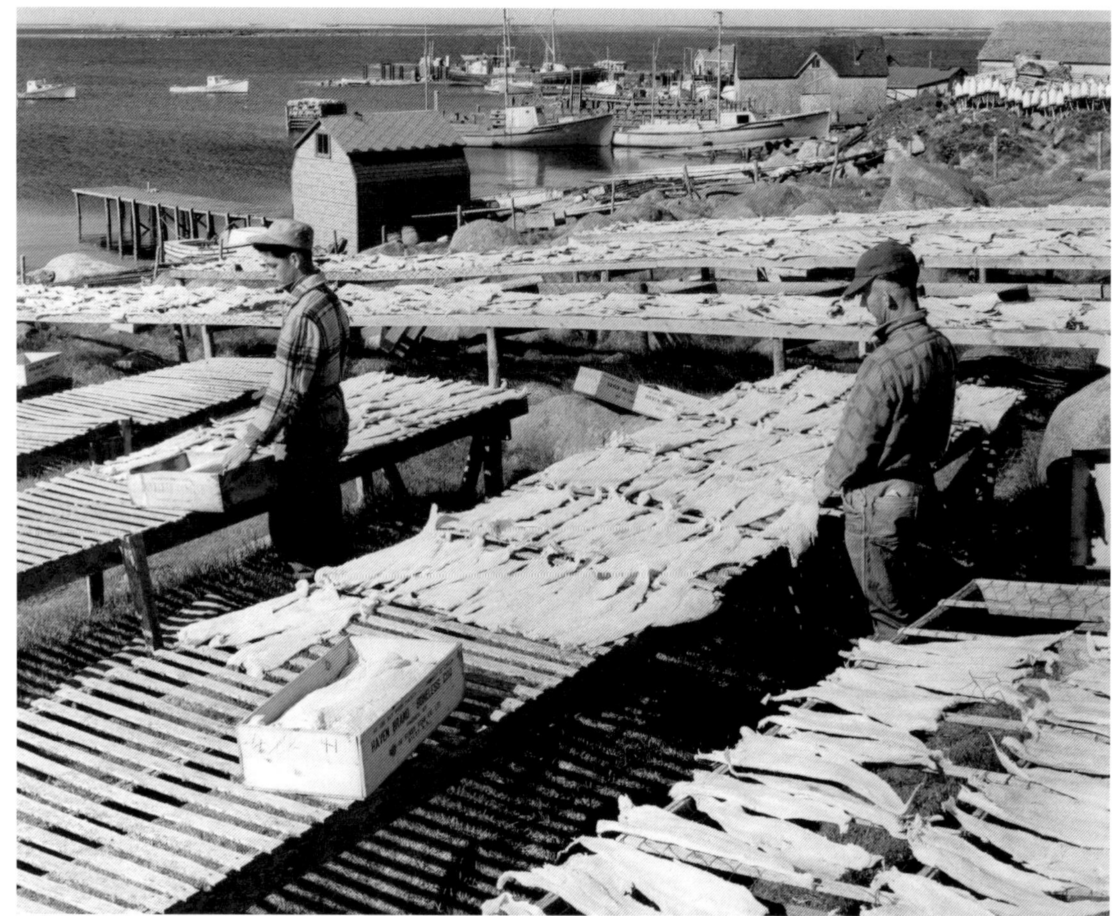

Drying cod, Clarke's Harbour, Nova Scotia, 1961. (Nova Scotia Archives)

Five hundred years ago, explorer and navigator John Cabot (c. 1450–1498) returned to England with fantastic stories about cod off the southeastern coast of Newfoundland, now called the Grand Banks. Schools of fish were said to be so thick they posed a navigational hazard, so plentiful they could be scooped up into baskets from a rowboat, and so dense, as Cabot put it, "that a person could walk across their backs." The news quickly spread and a cod fishery was established that helped feed the world for centuries.

Why was harvesting cod chosen over lobster? To begin, before the advent of refrigeration lobster had no shelf life; it had to be cooked and consumed within hours, whereas cod could be preserved almost indefinitely by salting. All that was needed to reconstitute cod's nutritious white flesh was a bit of fresh water, meaning it could be consumed anywhere and at any time of the year. Cod was especially valued as an easy and portable source of protein during long voyages at sea, land expeditions, and when going to battle. Besides the benefits of cod being so easy to catch and preserve, it was available in abundance. For these reasons, early settlers were eager to exploit the rich cod-fishing grounds of what is now called Atlantic Canada.

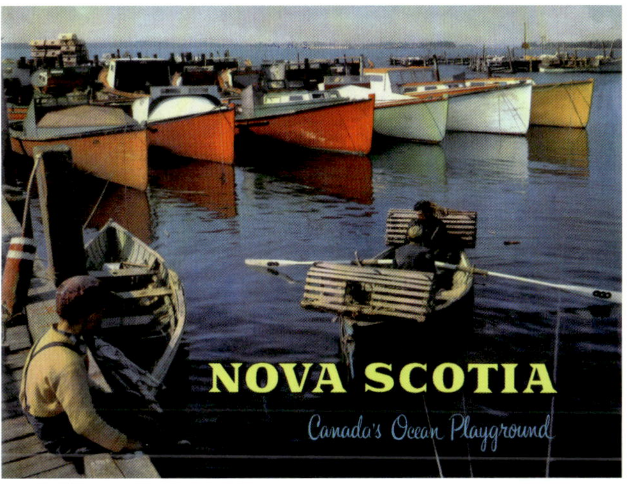

Nova Scotia Canada's Ocean Playground *tourism brochure, 1960. (Nova Scotia Archives)*

The first explorers to arrive here with the intent to exploit cod stocks and to settle the New World were the French. French Catholics had an important incentive for harvesting cod in those days: as a form of penance in reverence to Good Friday (the supposed day the crucifixion took place), they were required to abstain from red meat on Fridays. Fish on Fridays remained an important Catholic custom from the 1600s to the mid-1900s. Although less common now, this tradition remains in small pockets of French Catholic society within Europe and in Quebec.

Unappreciated and less versatile than cod, lobster was consumed as a lower-grade source of protein—and even then, only when cod reserves ran low and wild game or raised meat was scarce. The question is: why did codfish face near extinction while the lobster fishery remains viable and profitable?

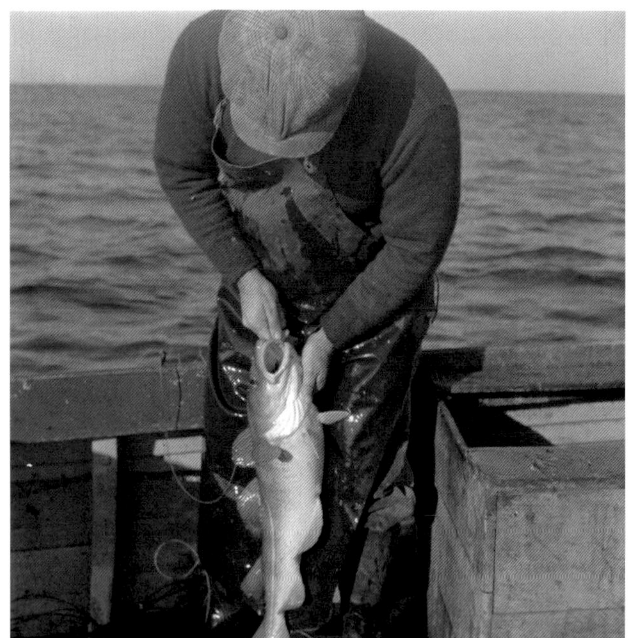

Not so long ago, in the "good ol' days," anybody could get out in their boat to catch as many cod as they wanted any time of the year. (Nova Scotia Archives)

The big difference between the cod and lobster fisheries is in the methods of harvest. Although the quest to build a better lobster trap continues, lobster-fishing methods have remained largely unchanged for over two hundred years. Sure, gone are the days of the oar-powered dory loaded down with traps; nevertheless, even with the introduction of the gasoline-fuelled boats of the early 1900s, and the more recent hydraulic-trap lifts, radar, GPS, and depth-sounders, not much has changed in lobster trap design and no sophisticated electronics can locate the reclusive, bottom-dwelling creature.

Lobsters were originally harvested by spearing or by gaffing. Both involved a sharp metal tip at the end of a long pole. The spear punctured the lobster from the top of its carapace. The gaff was a large hook inserted into the lobster's abdomen from its side in order to get the creature out of the water. To lure lobster close to shore, harvesters would reserve old cod heads for bait, and toss them in the water at night. Fishers would then set out the following morning to collect the lobsters. It was easy pickings. Depending on the tide and the shore's topography, lobster could be harvested from a dory or on foot in knee-high water. But these body-piercing methods were cruel. Lobster caught this way died in a matter of minutes and spoiled quickly. A damaged lobster is unappetizing and holds little monetary value. Since a live, whole lobster brought in a much better price, trapping lobster became the harvest method of preference.

In large part, we have the low-tech lobster trap to thank for the success of the fishery. Unlike fish, which typically travel in large schools, lobsters are territorial and spread themselves out across the ocean floor. The most advanced technology in the world cannot, could not, and will never be able to locate and harvest lobster any better than the inefficient, centuries-old traditional lobster trap. You could say that the lowly crustacean has unintentionally outsmarted advances in fishing technology. Dropping one trap at a time is the only way to catch the bottom-dwelling scavenger (forget snorkelling: that would be poaching, and fines are stiff).

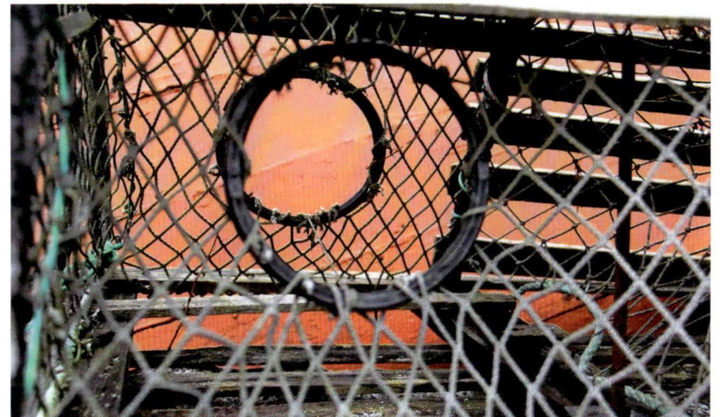

Loops on both sides of the trap guide the lobster into the "kitchen" where the bait is located.

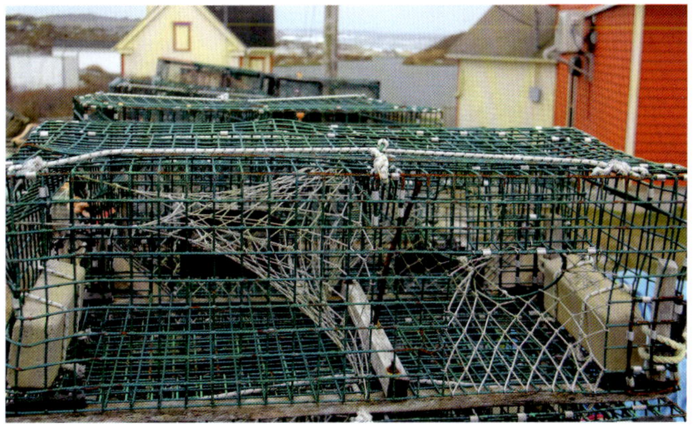

Once the lobster crawls through the mesh funnel into the "parlour," it is impossible to turn back.

A lobster trap is nothing more than a submerged cage with two parts: a "kitchen" and a "parlour." Two loop entrances called "doors" entice the lobster into the baited "kitchen." It is much easier to enter these doors than it is to exit them. Looking for a way out, the lobster is led through a mesh funnel into the "parlour" from which it cannot escape. The less adventurous lobster that stay in the "kitchen" have a better chance of escaping back out the loop doors.

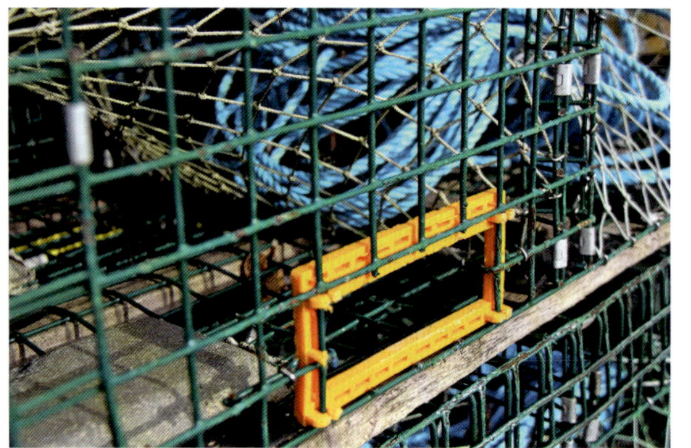

Trap exit vents are an important conservation measure.

Of all the lobster crawling the sea floor, only about 10 per cent will enter a trap and only 6 per cent are actually caught. Mandatory "exit vents" (1 3/4" x 4") allow juvenile lobsters and other bycatch species, like flounder, starfish, and sea urchins, to escape. Another 4 per cent of lobsters caught in a trap will be either undersized or egg-bearing females that must, by law, be returned to sea unharmed. So in the end, only 2 per cent of the estimated lobster population will make it to your plate. The inefficiency of the trap has inadvertently prevented lobster from being overfished.

The traditional method of catching cod, on the other hand, was with a jigger. Cod are predatory fish with a ferocious appetite, and will bite at anything resembling a small, moving fish. Though jiggers vary immensely in design, they all work on the same basic principle: The lure—a shiny, fake fish, usually made of metal—and hook are lowered into the water with a line and yanked up and down briskly. As the codfish attempts to bite at the lure, it gets hooked, usually by the mouth, and hauled up into the boat. Though a labour-intensive method of fishing, jigging was sustainable. This traditional method of catching cod had an unremarkable effect on cod stocks.

Unfortunately, advances in technology turned the one-hook-for-one-man tradition upside down in no time. The increasing use of high-tech gadgetry and the introduction of large factory ships in the 1960s suddenly made it possible to locate and net entire schools of cod (including immense quantities of bycatch "waste"). Lucrative codfish landings enticed more and more people into the fishery "boom." In Newfoundland and Labrador in particular, everybody wanted in on the game. Plywood fish processing plants popped up in just about every cove. Times were good. Coastal communities thrived and a lax attitude took hold. The fishery was left largely unregulated and poorly managed. More boats than ever before were out on the water scooping up stadium-sized nets loaded with cod. These mechanically operated nets were capable of swallowing massive amounts of multiple generations of the once-abundant fish. By the 1960s, 800,000 tons of Atlantic cod were being fished annually. By 1975, catches were in rapid decline with yields of less than 300,000 tons. A pattern of reproduction and a food chain that had evolved over millions of years had been irreversibly toppled. Greed, mismanagement—and, some would venture to say, ignorance—are what caused the great North Atlantic Cod Collapse.

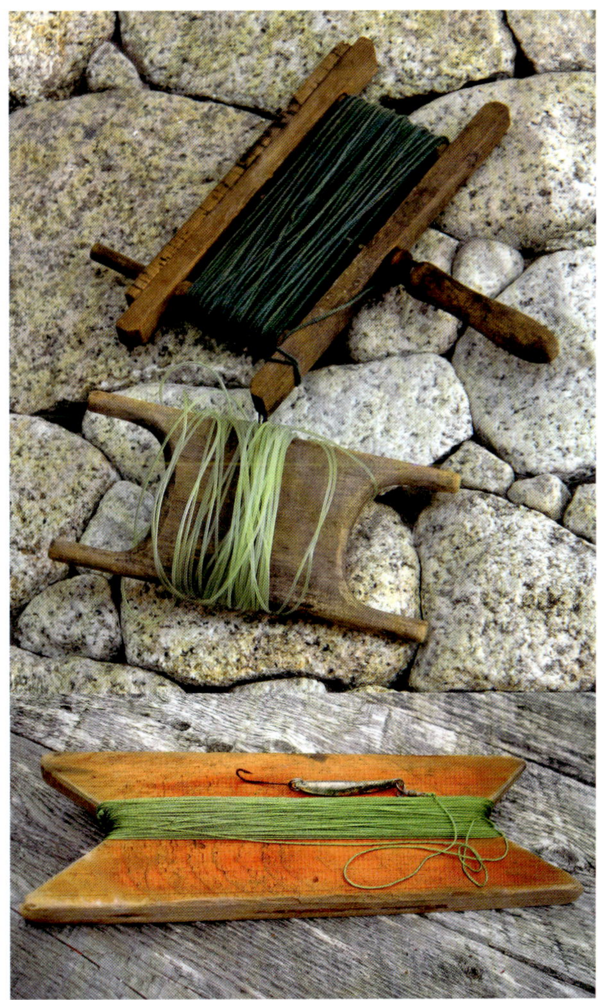

Turn-of-the-century "cod jiggers" caught just one fish at a time.

By 1992 cod catches on Canada's Grand Banks off the coast of Newfoundland and Nova Scotia plummeted to 1 per cent of the 800,000 tons enjoyed annually in the 1960s and early '70s. It was determined that cod stocks had become depleted; the bust hit, and it hit very hard.

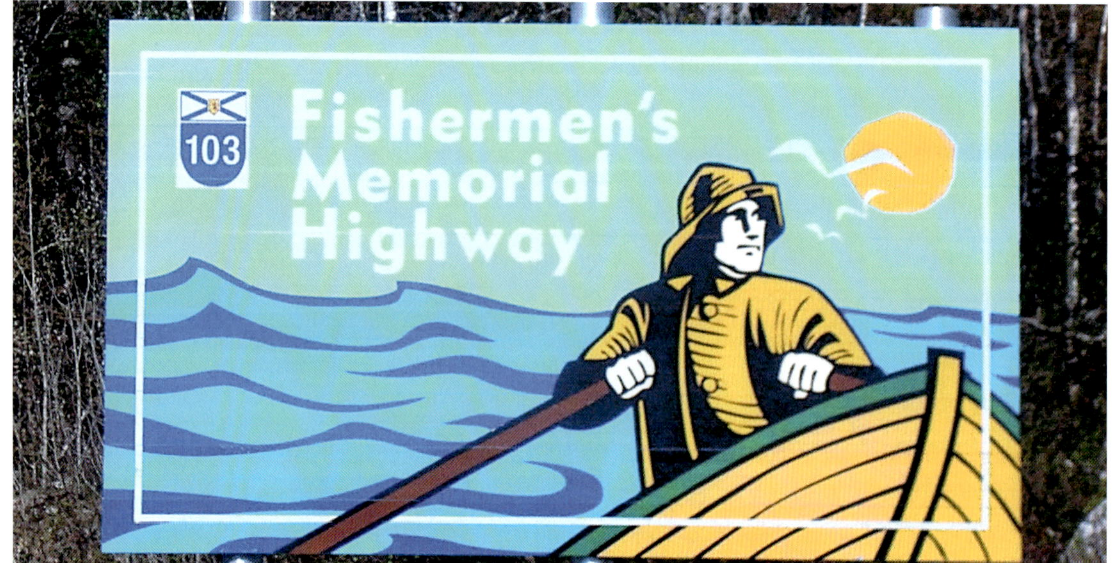

Nova Scotia's Highway 103 on the South Shore commemorates lives lost at sea, past and present.

The northern Atlantic cod was abruptly declared endangered. Afraid that cod stocks would disappear completely, the Canadian government put a moratorium in place in July 1992 to allow Atlantic cod populations to replenish. Over four hundred coastal communities that had been built on the cod fishery boom were left financially devastated and decimated. A way of life enjoyed for generations disappeared overnight. The moratorium put about thirty thousand people—12 per cent of Newfoundland and Labrador's labour force—out of work. It triggered the largest mass layoff in Canadian history. This resulted in a heavy reliance on government aid and outmigration. In the ten years following the moratorium, the province's population dropped by 10 per cent.

Today, the prized fish still has not returned to the numbers expected, despite the moratorium of twenty-five years ago. Cod size has also dramatically reduced. Overfishing the larger and most robust fish left behind a poor genetic stock. Lobster populationsn, on the other hand, remains healthy, but they too are much smaller than they were historically. The reason remains unclear, but the fact that they are heavily harvested cannot be ruled out. Were echolocation technology as useful in locating lobster as it is for schools of fish, lobster would probably have landed in the same unfortunate predicament as cod. (Many would blame seals, but they were never an issue until humans took a bite out of the balance of nature.) Most scientists suggest that the Atlantic cod will likely never recover completely. This was a hard and costly lesson learned. The only good news to come out of this sad saga is that the fishing industry now acknowledges the finite nature of the oceans. Atlantic Canada's cod fishery collapse has informed how the region's fisheries are managed today, including lobster.

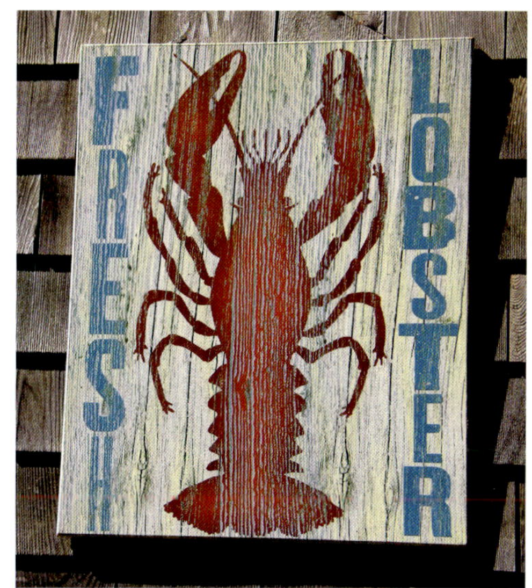

The live lobster market is a relatively new phenomenon.

The moratorium on commercial cod fishing remains in place with the exception of a few small, highly regulated experimental projects. One that proves promising is the use of individual "cod pots": semi-submerged netted cages. Fogo Island, Newfoundland, fishers have found this method to work very well; only market-sized cod get trapped. They can remain live and healthy for several days until the pot is hauled up. There is no carnage waste. Any bycatch is released live and unharmed back to sea. The fish is sold at a premium as a high-quality seafood product, mostly to restaurants. Fresh cod is considered a specialty in seafood markets. Salted cod flakes have become a thing of the past. Should the cod moratorium someday be lifted, this approach to fishing holds promise. However, it would be a very expensive, and thus prohibitive, fishery.

For now, the health of the lobster fishery hinges on stewardship based on what is known about the crustacean's life cycle and fluctuations in population. While some feel the fishery

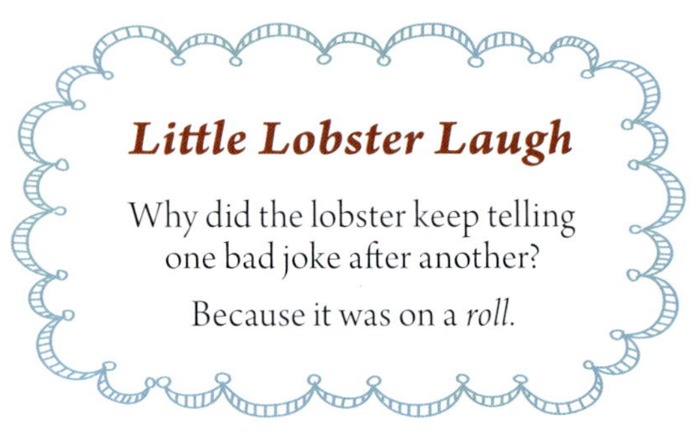

is regulated to death, most are in favour of how it is managed and cooperate happily. There will always be a few individuals who disagree with whatever is proposed, but protests mostly revolve around the issue of lobster pricing—the slippery slope of supply and demand.

The Department of Fisheries and Oceans Canada (DFO) works closely with the lobster-fishing industry to create conservation measures that meet the needs of each particular fishery. It is well understood that harvester input into fishery regulations is always a preferred solution to government-imposed laws in an industry that is constantly evolving. A team of industry experts comprised of oceanographers, harvesters, processors, and DFO officials draft lobster-fishing regulations. These include limits on the number of licenses and traps issued, trap-design specifications, fishing season lengths, catch quotas, and lobster size restrictions. The team is also responsible for spreading the fishing season over the various fishing zones (see map on page 90).

The elusive nature of lobster and a few commonsensical regulations, then, are what keep Canadian lobster stocks in such good shape. Atlantic Canada's lobster fishery is the most profitable in the world. It is reputed to be the best managed and most sustainable seafood industry in both export and domestic sales. The Atlantic provinces are famous for fresh seafood, but lobster is the region's most important fishery with landings worth $620 million a year and resulting in $1.5 billion in live lobster and lobster product exports. Atlantic Canada can boast being the leading exporter of cold water–caught lobster in the world. The future of white-fin tuna, wild Atlantic salmon, halibut, and various whales and sharks is uncertain and may very well go the way of the cod, but with continued good fishing practices and conservation measures, the future of the lobster fishery holds promise for many generations to come.

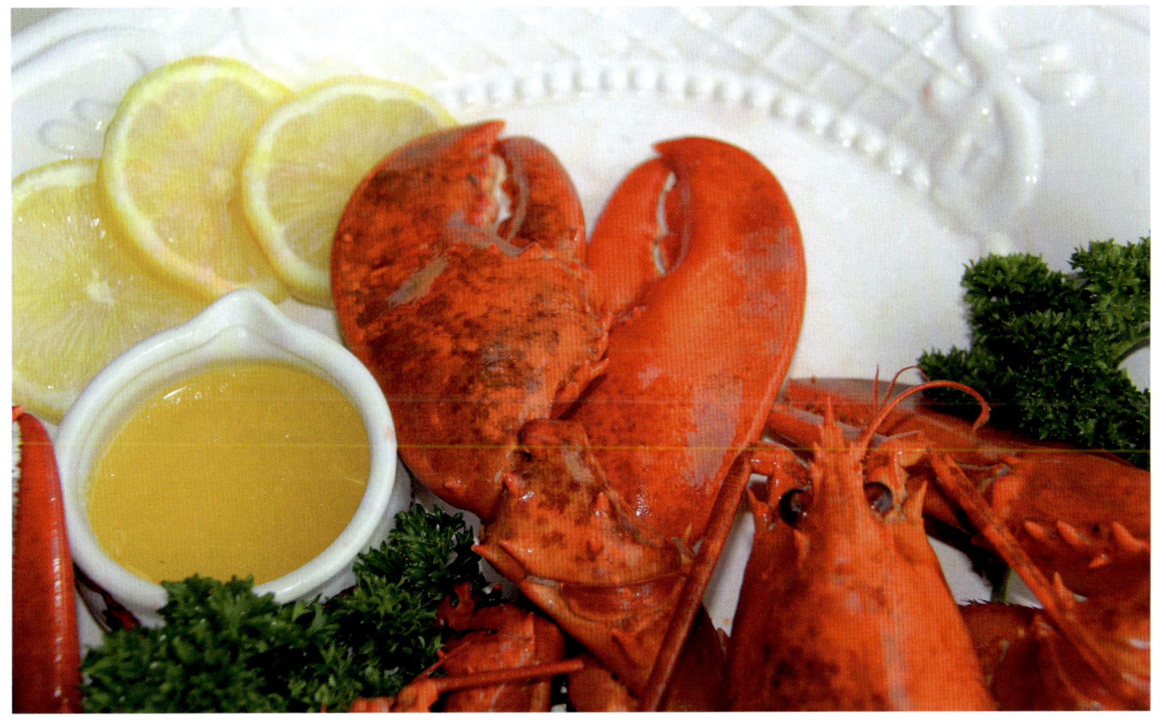

Whole lobster in the shell can be cooked and enjoyed much more affordably at home rather than in a restaurant.

From Cockroach of the Sea to Luxury Food

People did consume lobster before it gained popularity, but not happily. As mentioned earlier, lobster was only eaten out of necessity; in fact there are many accounts of lobster being harvested to feed the marginalized. During the Colonial era (1600s to late 1800s) in what is today called New England, lobsters were fed mostly to servants, slaves, and prisoners. Just two or three generations ago, Atlantic Canadian children sent to school with lobster for lunch were considered poor; those with a bologna sandwich were the privileged. Lobster was so underappreciated, it was sold as one of the first canned pet food products of the early 1900s: Atlantic Brand's Maine Lobster Pet Food for Pampered Pets. However, as canning and refrigeration methods improved, so did the quality of preserved lobster for human consumption.

Lobster Processing in Canada

Traditionally, before the advent of the live-lobster market, lobster were cooked as soon as they were landed. What wasn't to be consumed in the next few days, was canned. Today, the live-lobster market represents about half of all lobster landings; the other half is processed. Legal-sized small lobster ("canners") are considered too small to be consumed as whole in the shell. These and flawed lobster ("culls"), those with a deformity or missing a claw, are sent live to processing plants to be transformed into lobster products.

The lobster-canning industry has a long history in Canada. It began in Miramichi, New Brunswick, in the 1850s and quickly spread throughout the Atlantic provinces. The 1970s brought significant advances in refrigeration technology. Lobster meat, raw or cooked, freezes exceptionally well and is of superior quality to traditional canned lobster (which is now found in the freezer section of supermarkets). Frozen lobster products gradually replaced canned lobster.

Preserved lobster takes many forms now. New and innovative ways of packaging frozen raw and pre-cooked lobster opened up a wealth of food-service markets. Specialty frozen lobster products are available all over the world to customer specification: tails, claws, whole lobster in the shell, in gourmet sauces, lobster spreads, cocktail-style lobster, and so on.

The United States, followed by Europe and Asia, is the greatest consumer of frozen lobster products (for restaurants, the cruise ship industry, casinos, etc.) New Brunswick is Canada's main lobster-processing province (25,000 metric tons per year), followed by Prince Edward Island (14,000 metric tons), Quebec, and Nova Scotia. Lobster, and lobster products, remains Canada's top export species at a value of $1.5 billion annually. In March 2014, Fisheries and Oceans Canada reported $4.9 billion in Canadian exports of fish and seafood products (including lobster), an increase of $517 million from 2013.

Packing lobster in factory near Cheticamp, 1959. (Nova Scotia Archives)

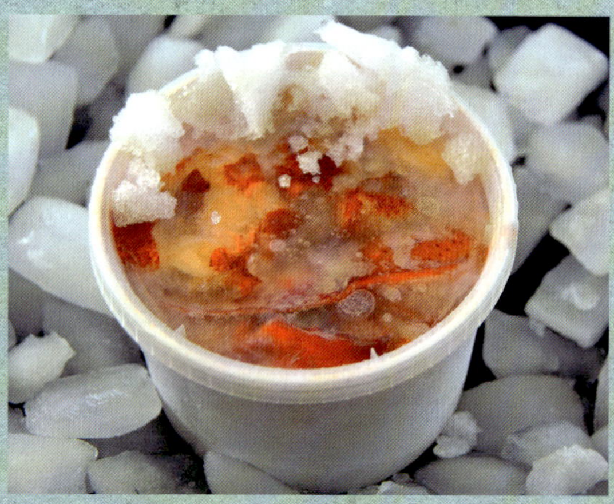

The quality of lobster meat preserved this way is just as versatile and delicious without the mess.

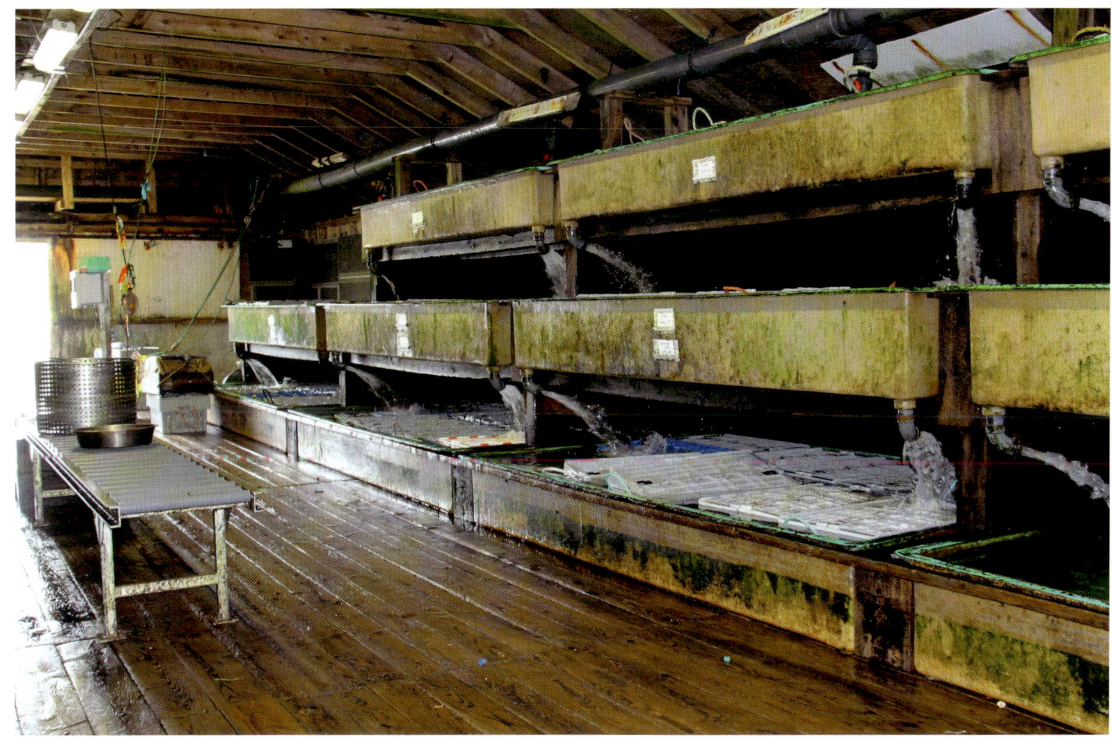

Lobster can be kept for as long as six months in lobster pound facilities but rarely stay that long. They can then be shipped live to markets all over the world at any time of year.

The live-lobster market developed in the 1870s as railroad tracks began linking the outskirts of coastal Maine with booming cities like Boston, New York, and beyond. The first shipment of live lobster packed on ice was done by rail. American entrepreneur and newspaper tycoon William Randolph Hearst ordered it for a private dinner party in Colorado. He wanted to impress his guests with something new and exotic. Caterers cooked and served the Maine lobster on the spot.

The little-known crustacean from the North Atlantic dazzled those who had it for the first time. Very few foods were so brilliantly red and delicious as this novelty from the sea. Lobster became an instant hit, but was restricted only to those willing and able to pay a premium to have it shipped live on ice from Maine, via rail or bus, to nearby states. We can thank the

The Lobster Pound Phenomenon

Lobsters are highly perishable and market demand fluctuates. Unlike fish, lobster cannot be sold for human consumption once they have died; they must be sold processed or live. Lobster can live up to two days out of water as long as they are kept cold, damp, and out of sunlight. A lobster harvester's first priority is to get the catch to the nearest lobster pound as soon as possible.

A lobster pound is a large commercial indoor tank where temperature-regulated seawater is continuously pumped in. There, lobsters can be kept alive for several months, making them available fresh year-round. Operators specialize in the shipment of live lobster locally and to major markets all over the world at any time of the year. Some lobster pounds have several tanks where lobsters are sorted by weight, while some smaller, local lobster pounds supply large seafood distribution companies such as Clearwater Seafoods Ltd. in Bedford, Nova Scotia.

Harvesters deliver their catch live to the pound in crates or mesh cages, where workers sort each individual lobster by weight to customer specifications.

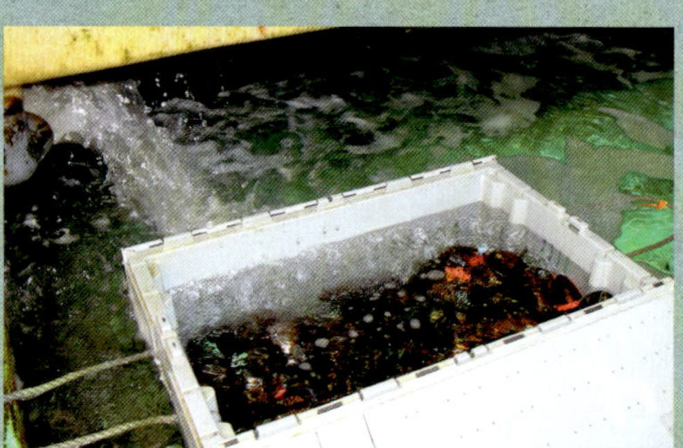

Water is pumped directly from the sea, keeping pound tanks oxygenated and the prized crustaceans healthy for the live market.

rapid spike in the live lobster market for the advent of lobster pounds and advancements in transportation. Together, they changed the lobster's reputation and destiny forever.

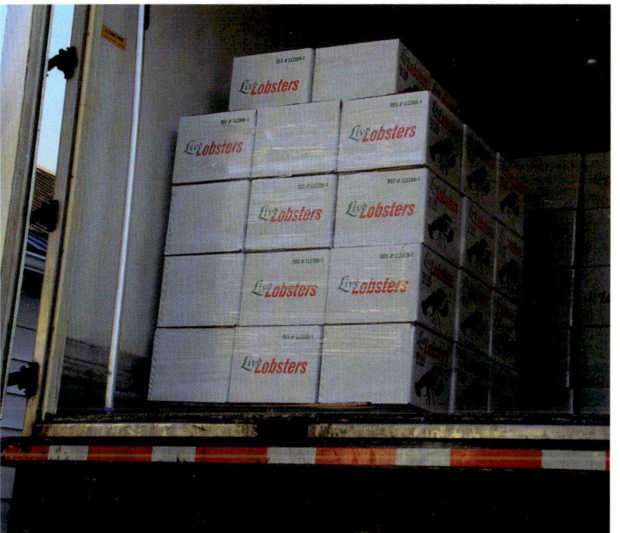

A load of live lobster packed in ice in a refrigerated transport truck from Ryers Lobster Retail is destined for various markets overseas via a cargo-jet aircraft out of Halifax Stanfield International Airport.

The demand for whole lobster, in the shell and served steaming hot, spread quickly. By the mid-1900s, Maine lobster fishers could hardly keep up. Lobster populations began to suffer as a result, and harvesters had to adopt strict measures to ensure a sustainable fishery. When Maine stocks got low, Canadian-caught lobsters filled the gap, and still do to this day.

Globalization and easy access to aviation paved the way for the reliable, next-day delivery of live lobster on ice to international markets. With so many other fisheries in trouble, the demand for lobster has grown exponentially and continues to expand. In the grand scheme of things, lobster's evolution to delicacy from such humble beginnings occurred at lightning speed. The lust for lobster at home and abroad shows no signs of slowing down.

Water Temperature Dictates Lobster Quality

Lobster thrive in the cold waters of the Atlantic Ocean. This explains why Canadian-caught lobster is of premium quality. Lobster are cold-blooded creatures that prefer cold water, but the animal can freeze to death too. Those areas with a winter fishery have particular challenges. Lobster can die from exposure in a matter of minutes in subzero temperatures.

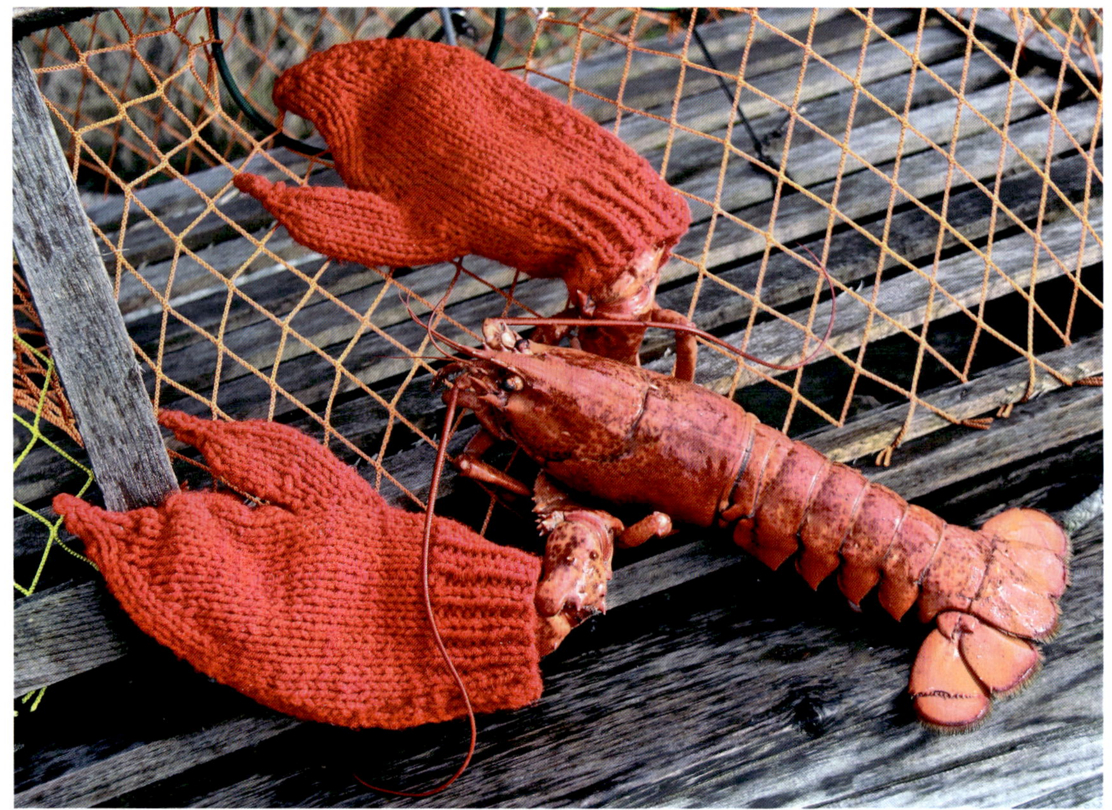

These hand-knit lobster-claw mittens are an ode to Atlantic cold-water lobster: the best in the world.

Lobster are at the greatest risk of freezing to death when they are landed. Time is of essence on a cold winter's day! Luckily, fishers have a brilliant way around this.

To prevent lobsters from freezing on the boat, once taken out of the trap and banded (placing a rubber band over the claws to keep them shut), they are quickly placed in tall, wide barrels of seawater. Now there is a risk of them drowning. Yes, lobsters can die from lack of oxygen if the water gets stale. To keep the water oxygenated, fishers add fresh seawater to the barrel about once every hour to introduce life-sustaining oxygen to the lobsters inside. As long as new seawater is continually added and ice doesn't form, lobster will remain alive and healthy. It's a lot to coordinate on a freezing, bumpy, wet ocean ride. Once back to the wharf, if the temperature is much below 0 degrees Celsius, it's not worth the risk of loading the catch on the truck for the lobster pound—even if it's just a five-minute drive. Rather, the lobster are kept in the sea, inside secured cages attached to the wharf.

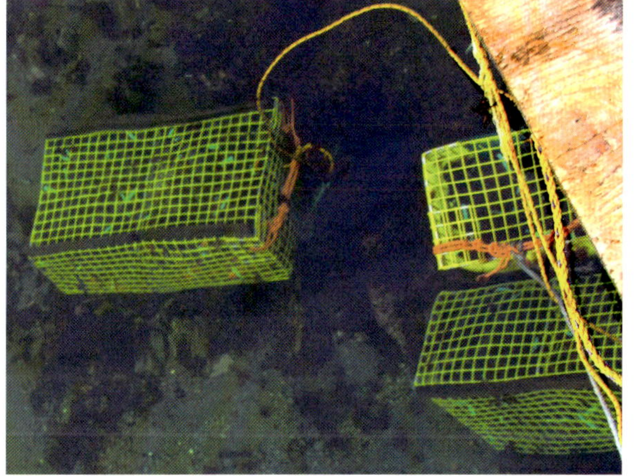

Cages full of banded lobster are kept under water until the weather is more favourable for transferring them from wharf to truck to pound.

Freezing to death isn't the only issue facing landed lobster: warm conditions can kill too. For this reason, climate change is of concern to the future of the Atlantic Canadian lobster fishery. There are already signs of lobster being affected. American scientists and the fishing industry warn of dire predictions for Maine lobster, due to climate change. This decade, surface and shallow waters in the Gulf of Maine have hit an all-time high of 20 degrees Celsius (10 degrees Celsius is the warmest temperature lobsters can endure). As water temperatures keep rising, so does northern lobster migration. It is now evident that Maine lobster stocks are

"We were fishing overnight. We had caught about two thousand pounds of lobster. They were stored in an insulated crate with a tarp on top. The temperature dropped to -11 degrees [Celsius] that night and slush was building on the deck of the boat. Now, in the daytime with the sun out, that temperature wouldn't be a problem. But it was night. Once we got the lobsters to the pound, those ones that were at the very bottom of the crate didn't make it. Dead lobsters are not marketable in any shape or form. From now on, as a precaution, we keep the lobsters in a shelter when it's that cold so this doesn't happen again."
–Roy Fralick, lobster fisher, Peggys Cove, Nova Scotia

gradually making their way toward the Atlantic Provinces. While this is good news to Canadians at the present time, oceanographers predict that as seawater continues to warm, lobster will inevitably move even further north of the Atlantic provinces.

Further to the northward migration of Maine lobster, when exposed to warm waters the crustacean becomes more susceptible to pathogens. We've recently seen the emergence of shell diseases caused by bacterial infections. In these warming conditions, lobster also have to expend more energy to "breathe," which makes getting around cumbersome. Lobster can become stressed, leading to self-amputation or the diminishment of one or both claws. They moult earlier and it takes longer for their new shells to harden. Once harvested, stressed lobster don't remain alive as long. It is hypothesized that with rising temperatures, lobster will abandon the Gulf of Maine altogether.

Dean Ryer, of Ryers Lobster Retail in Indian Harbour, Nova Scotia, gives the following advice about proper water temperature:

> *Water temp is very important. Water temp usually above 10 degrees [Celsius] can promote commitment to moult and also causes stress for the lobsters that are held in large volumes in lobster pounds—less oxygen, for example. But on the positive side, warmer water makes the lobster more active in [its] natural habitat... looking for food or entering fishermen's traps, etc. Also, [being] held in warm water for a short time helps purge lobsters quicker... this is important for shipping them long distances, but [they] should be put in colder water for a time before shipping.*
>
> *Regardless of the global warming of our water, you must refrigerate the natural water coming into your plant locally from mid-May till late October. The preferred temp for shipping 24–48 hours is, in my opinion, 4–6 degrees [Celsius]. For storage long term (3–4 months), the temperature should be 2–3 degrees [Celsius].*

The term "purging" Dean refers to is significant. As with every other living creature, lobster, too, urinate and defecate. This is where lobster and dogs have something in common. Dogs, both male and female, use the scent of their urine to communicate with other dogs. They mark *where* they have been and *who* they are with their urine "signature." Unlike dogs, though, lobster have two bladders—both of which are located in the head. They relieve themselves through exits near their antennae. Lobster are a little more extreme than dogs, in that they meet and greet each other by peeing in each other's face! But then, there is a lot of water around so it doesn't have much lasting power.

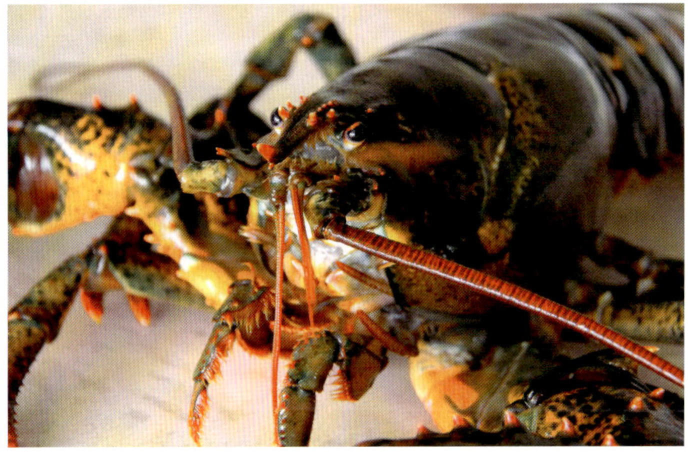

Although they don't exibit much expression, lobster are extremely temperature sensitive. The difference of a few degrees either way can be fatal to them.

As you can imagine, after having helped itself to a load of easy bait, a lobster is usually full when it is pulled out of a trap. If cooked and served before a purge, its insides on a plate wouldn't look very appetizing and the flavour of the meat inside would be affected. In order to prep lobsters for market, digestion must run its course so the lobster is cleared of residual stool and urine. The other reason for purging lobster is because when contained in a tank, the resulting ammonia can quickly reach toxic levels.

What does this have to do with water temperature? Lobster digestion is induced when briefly placed in warmer water. When lobsters arrive at the pound, they are first put in receiving tanks of seawater pumped directly from the ocean for two to three days for purging. Then they are moved to holding tanks of much colder refrigerated seawater until they are ready to be shipped off to markets all over the world.

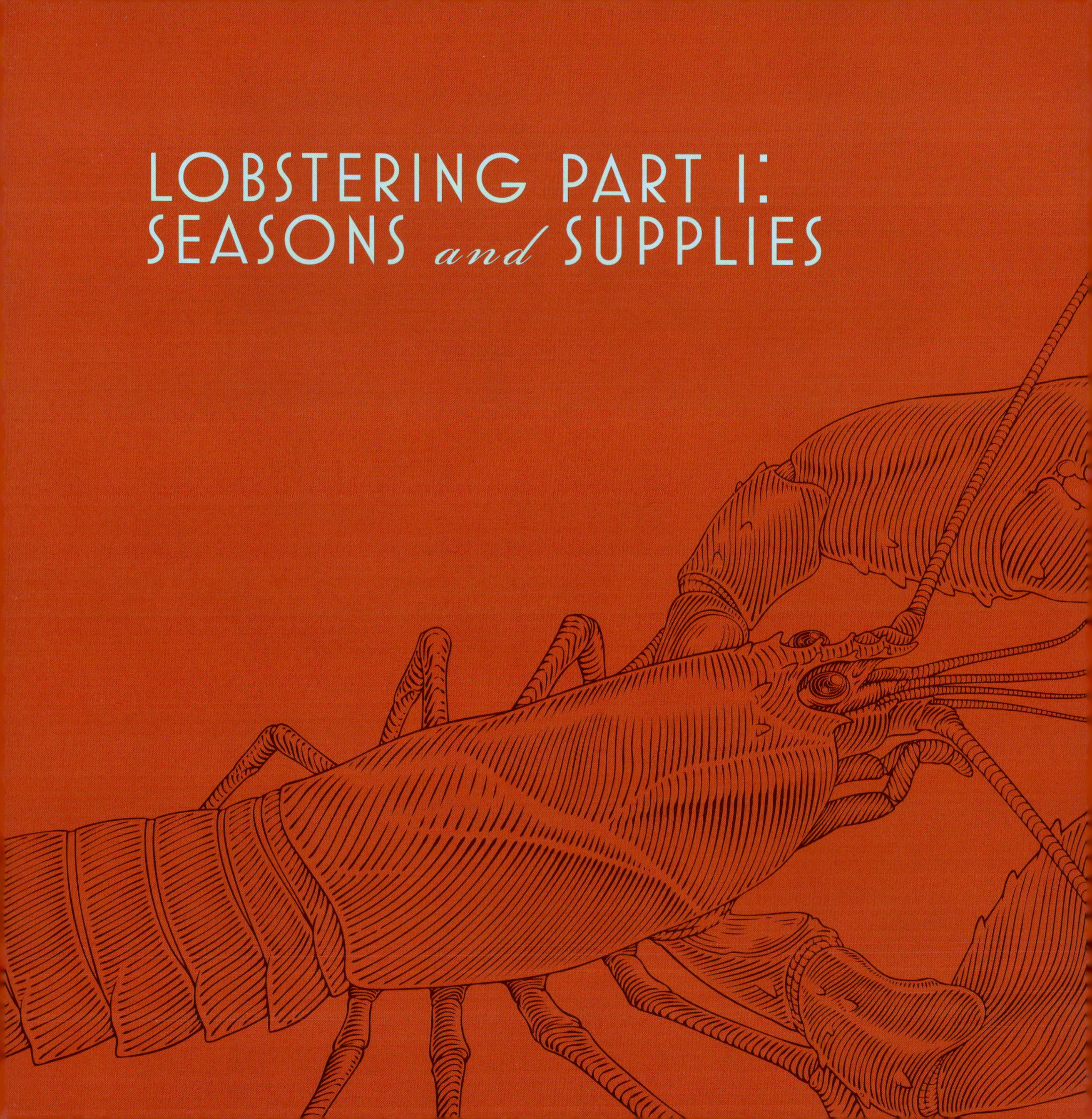
LOBSTERING PART I:
SEASONS and SUPPLIES

A Lobster Fisher's Work is Never Done

There is no such thing as weekends off in this business. The season is short and sea conditions are not always favourable. Lobster harvesters must maximize their time on the water in order to succeed. It's a hard workday that begins at about 4:45 A.M. when it's still dark out. Get some breakfast, pack a lunch, fill the water jug, get into the rubber gear and boots, load the truck with boxes of slimy thawed bait, top up the fuel cans, and head for the wharf. Once down the slippery ladder into the ol' boat, flick the lights on. If there's ice on the deck, out comes the pickling salt. Fire up the engine, grease the hydraulic trap lifter, and begin chopping the bait.

Why start the day so early? There is so much to do before setting out to sea, the earlier the better. Furthermore, lobsters are most active at night. The longer the traps are left under water, the better the chance of a lobster figuring out how to escape: income lost. Another reason to get at the trap early is because if two or more lobsters get in, they will fight, as they are fiercely territorial. This can result in injuries that compromise lobster value. Also, left in strong currents a trap can roll on the sea floor, damaging gear and injuring the catch.

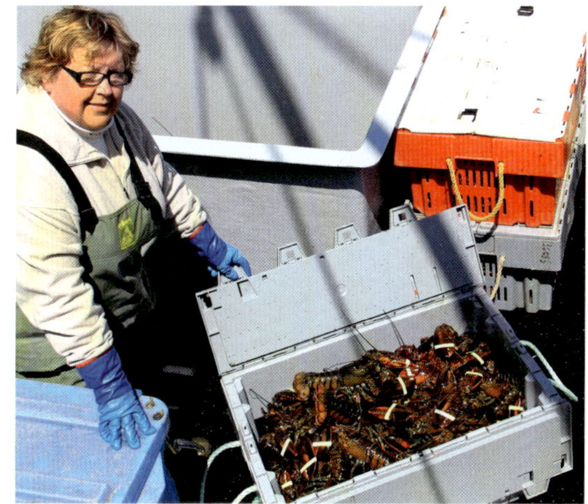

More and more women participate in the lobster fishery.

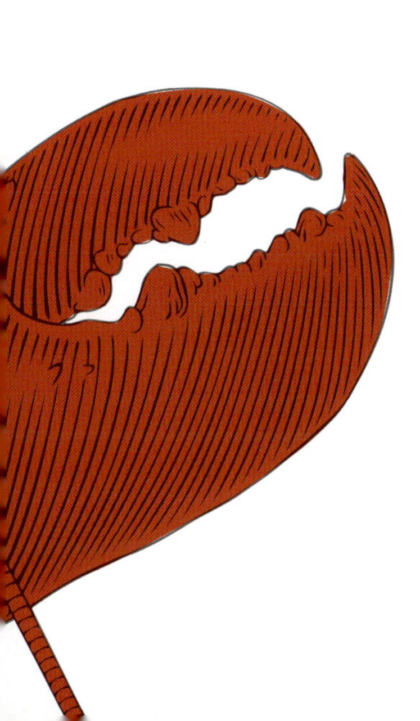

As with all things in life, the lobster fishery, too, has seasons. There's a prep season and a fishing season. The latter is a highly regulated period for an obvious reason: conservation. Gone are

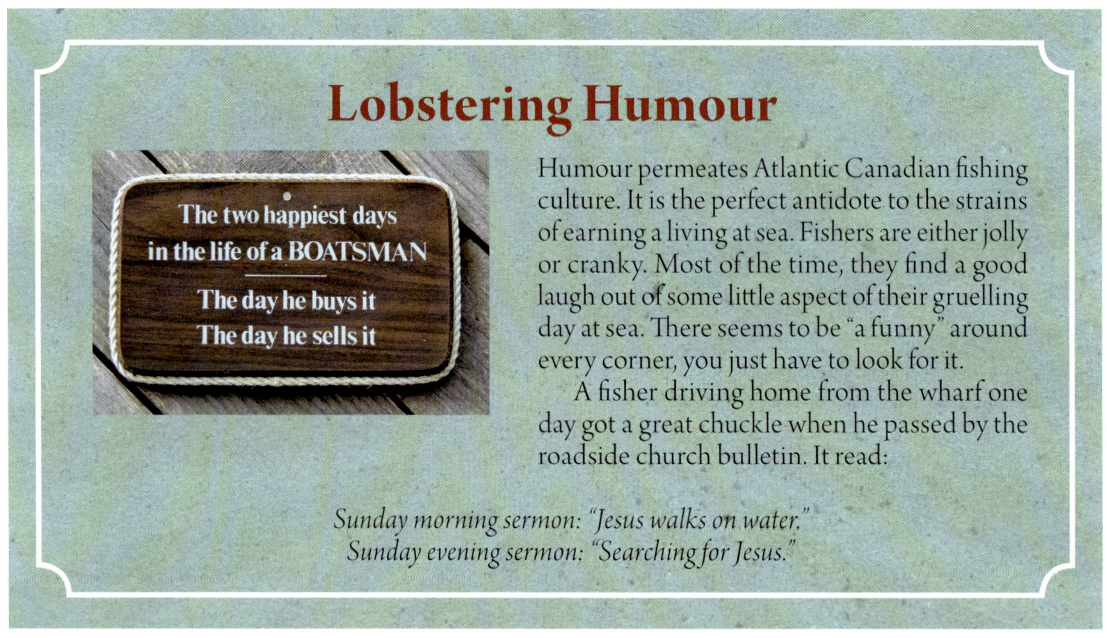

Lobstering Humour

Humour permeates Atlantic Canadian fishing culture. It is the perfect antidote to the strains of earning a living at sea. Fishers are either jolly or cranky. Most of the time, they find a good laugh out of some little aspect of their gruelling day at sea. There seems to be "a funny" around every corner, you just have to look for it.

A fisher driving home from the wharf one day got a great chuckle when he passed by the roadside church bulletin. It read:

Sunday morning sermon: "Jesus walks on water."
Sunday evening sermon: "Searching for Jesus."

the days of going out in your own boat to catch your next meal from the sea whenever you please. Licensed lobster fishers, for the most part, cooperate with Department of Fisheries and Oceans (DFO) regulations rather than risk a charge of poaching. Incidents of poaching are relatively rare in the Atlantic provinces. Most folks in these parts have a lot of respect for what goes into earning a livelihood fishing for lobster. A basic lobster-fishing license is about $250,000. And then there is the cost of the boat, electronic accessories, the traps, buoys, bait, and fuel. It's almost impossible for newcomers to get into the industry fresh except when licenses are passed on from a relative. For reasons listed above, the cost to getting started on your own is exorbitant.

And don't think for a moment that the off-season is vacation time. Many lobstermen opt to supplement their lobster income with other fishing licenses for marketable seafood products such as haddock, halibut, flounder, shrimp, or snow crab. Undersized fish and non-marketable species are sold as bait.

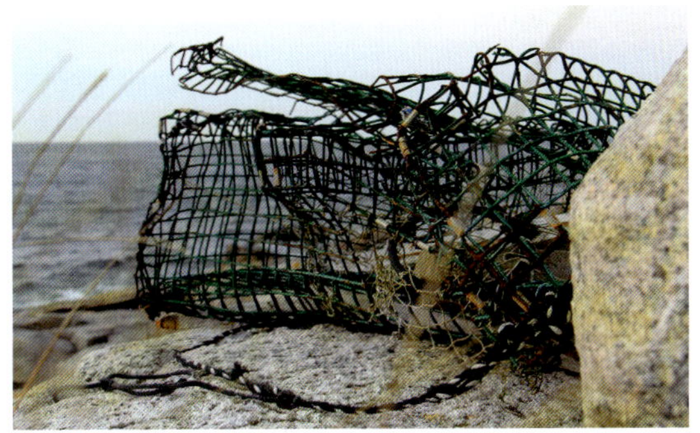

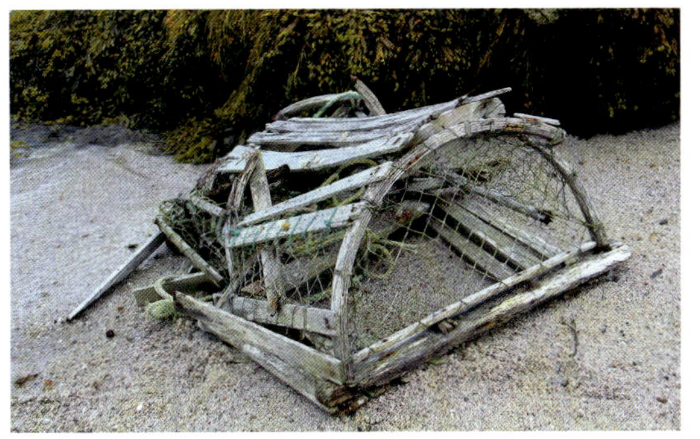

Storm-damaged traps are a common sight along Atlantic coasts.

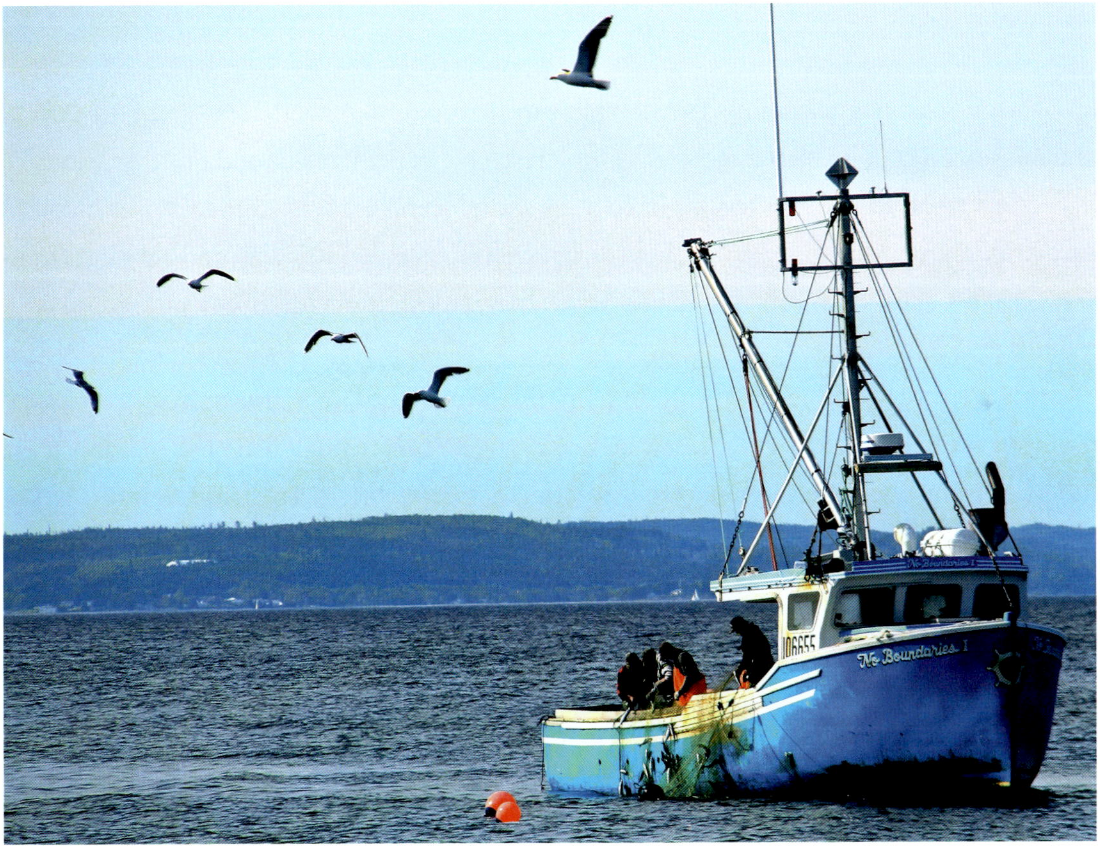

Lifting nets of mackerel destined for lobster bait.

"Yeah, me and my dad, Richard Hubley, have been fishing forever it seems. We each have our own boat now. We have a winter lobster fishery here in St. Margarets Bay. The weather can get bad at times, but the winter lobster is the best quality 'cause it's hard-shell. We catch a good ten thousand lobsters each the first week. After that, we only get half that amount the rest of the time till the lobster fishery closes. Those areas with a summer fishery get soft-shell lobsters. Those ones go to processing for canning or get frozen. Then, in the summer, we have the groundfish license. We catch mostly mackerel and haddock around here. We also go up to the Bay of Fundy for three days to fish halibut. Last week, we got one that was bigger than this worktable. It weighed a good 265 pounds. A massive fish!" –**Tim Hubley, lobster fisher, Modesty Cove, Nova Scotia**

The off-season is also a time to catch up: time to repair damaged traps, paint the buoys, and for general boat maintenance, including a fresh coat of paint inside and out. Some lobster harvesters resort to odd jobs to supplement the short lobster season's unpredictable income: carpentry, chopping wood, painting houses, drywall work, mowing lawns, and plowing driveways are just a few examples. No matter what, these are resourceful folks who find ways to make due in hard times.

When the economy is down, lobster demand lowers—and, as a result, so does the price. Going to sea to earn a living wage may be a risky roller coaster, but lobster fishers value their independence. During the open lobster season, they can work as much as ten hours a day. Time to quit is dictated by nightfall, cold, sea conditions, or when a backache tells them they've had enough for the day.

Shacks, Traps, and Superstitions

Most people's contact with the lobster fishery is from a distance: sightseeing quaint coastal communities where peaceful, weathered wharves jet out over sparkling, sheltered waters.

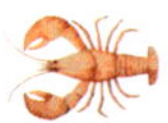

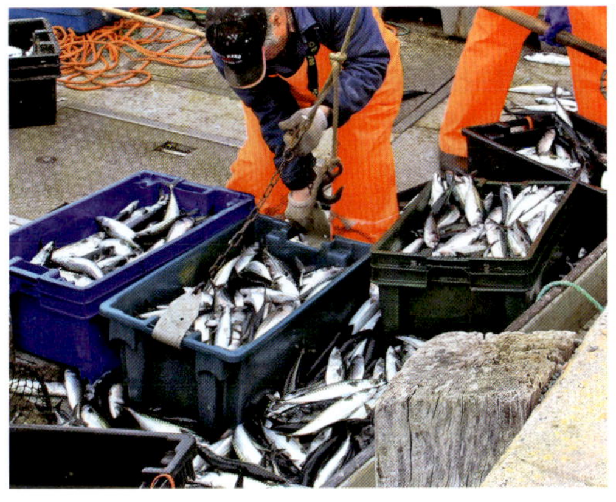

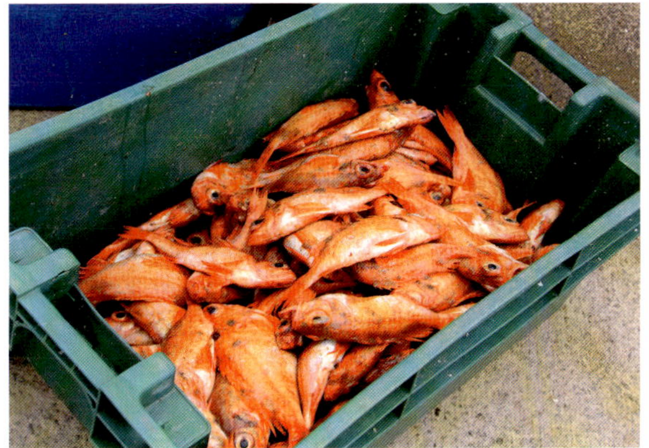

Bait preference depends on availability. Mackerel (left) and redfish (right) are abundant in waters off Nova Scotia's South Shore.

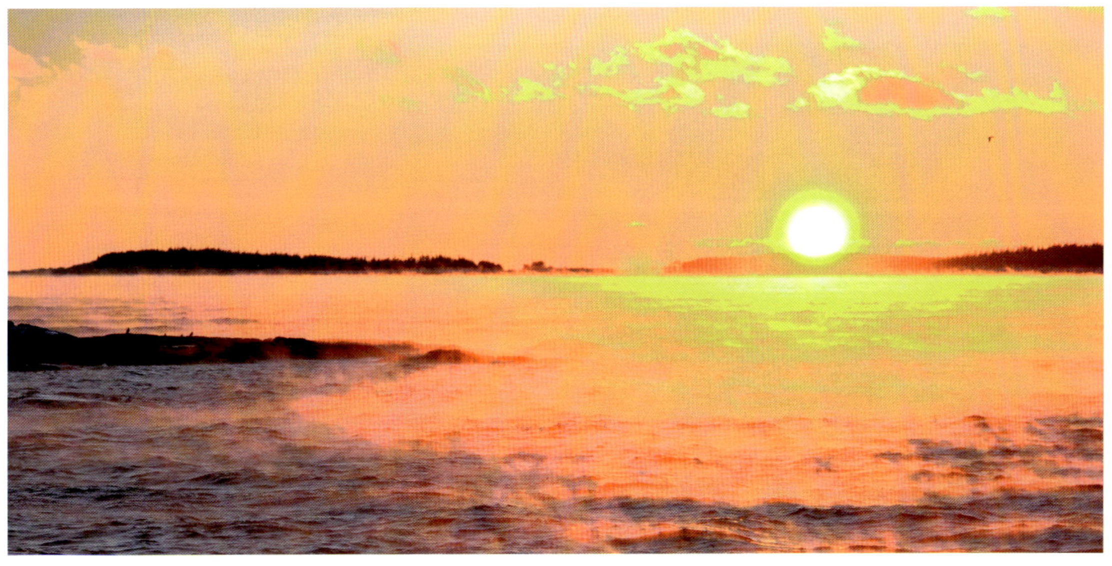

Winter sea fog means it's too cold for fishing.

Eye-popping colours from boats, traps, buoys, ropes, and floats make for great photographs and may very well bring an old sea shanty or two to mind. The fish stores, often called fishing shacks, are another remarkable cultural feature dotting coastal communities. They serve many important functions depending on the fishery. Those with a groundfish licenses use them for storing their gear, filleting, and salting fish. Many lobster harvesters keep their traps out of the weather in these sheds when the fishery is closed. They also use them to store tools needed to repair traps, paint buoys, or for fibreglassing an old wood dory.

Wharf-side sheds have been used to build, cork, and paint boats. They also serve as a place to store bait. But most of all, they are a gathering place to share stories about The Big One That Got Away and to reinforce a few superstitions. Some of these are personal, others are local, and a few are kept just between husband and wife. Superstitions can differ from harbour to harbour, but many are widespread.

In some cases, superstitions directly affect the industry. It is said that change was slow when it comes to modern lobster trap design, despite the acclaimed advantages. The old adage

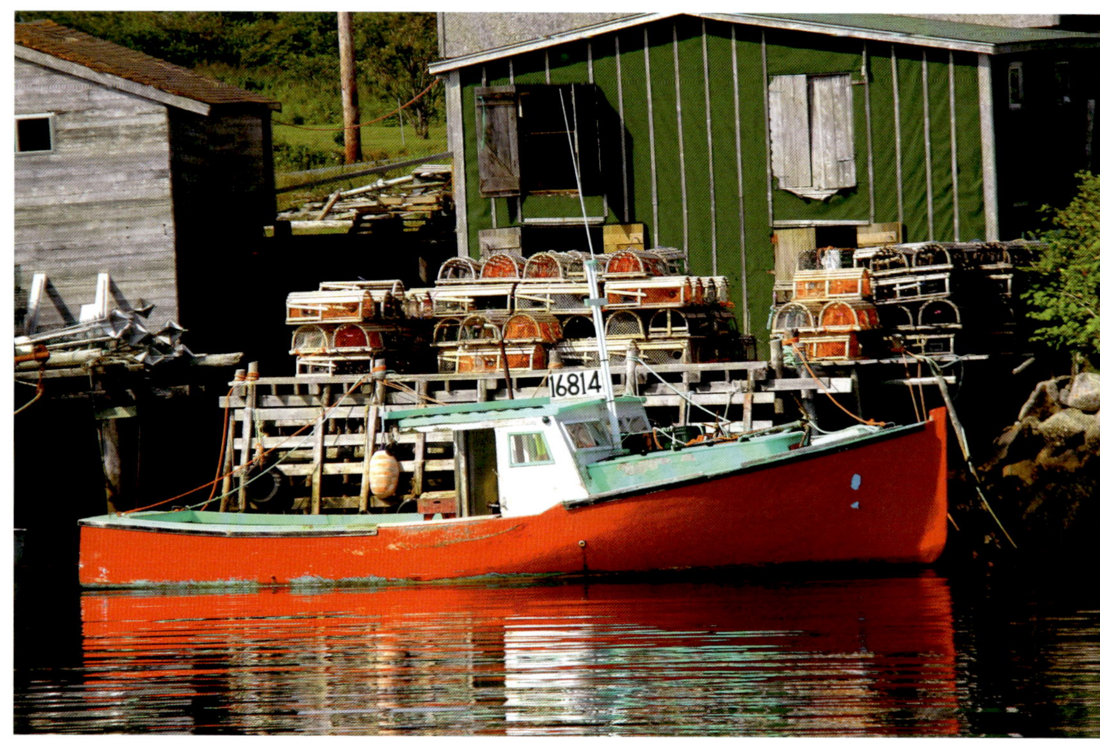

Fish stores are very utilitarian; not much goes into their appearance, which is part of their charm.

Fibreglassing

To get a few more years out of an old, leaky wooden boat, many fishers turn to what is called "fibreglassing." First, they weigh the cost of buying or rebuilding a boat against "preserving" an old wooden boat in resin. It is often deemed cheaper to keep the wooden boat seaworthy by sealing it in fibreglass cloth over resin and then coating it in resin again. The authentic wooden appearance and value of the vessel is lost in this procedure. The glossy-brown mesh restoration is purely practical, but the boat's life has been extended by at least another decade.

If it ain't broke, don't fix it led to fears that using traps built without the traditional half-circle shape and made of materials other than wood would be disastrous. Fishers tend to hold on tightly to conventions, habits, and routines.

Despite the contagious nature of these warnings about trap changes, a few adventurous younger fishers were willing to break tradition and show that nothing bad would happen with the use of stackable wire mesh traps. The only downside to the design is that when one of these traps breaks free from the buoy line (and it happens) it remains on the sea floor for decades. For this reason, as of 2013 each trap must have biodegradable exits in order to minimize harm to sea

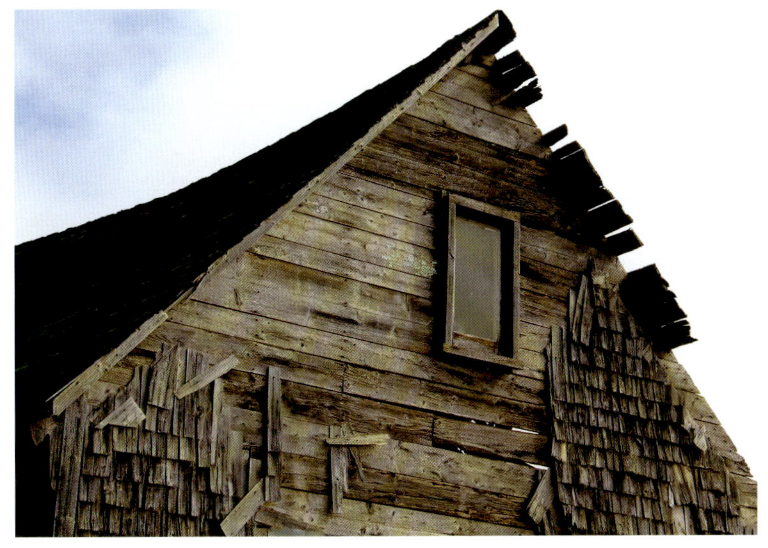

This old fishing shed built over a century ago is still in use. What stories it could tell...

Superstitions on the Sea

Here are a few common superstitions held among fishers:

- Never whistle a tune on the boat or it will bring on a gale.
- Don't untie a boat from its mooring until the engine is running.
- Never say "good luck" to a fisher heading out to sea. (It will result in a jinx.)
- Changing a boat's name is bad luck. (Famous renamed vessels have sunk.)
- No bananas allowed on-board. (Their peelings are slippery.)
- Always give a boat a female name. (Only "she" can appease the sea gods.)
- Never eat pork on the boat. (Pigs can't swim, so you will drown.)
- Buoys painted blue or green bring on a curse. (Well...they are hard to see.)
- Red sky at night, sailor's delight; red sky in the morning, sailors take warning.
- It's bad luck for a fisher to learn how to swim. (The ability to swim is an indication that your boat will surely sink at some point.)

Some superstitions, though amusing, seem to be grounded in a bit of common sense. They live on because they give a sense of control over what is often perceived to be beyond one's control out at sea. For instance, generally people who earn a living at sea can't swim—unless, it is said, they were born with webbed feet. Though they grew up by the water, the sea was perceived mostly as a place of work rather than one of leisure. That many older coastal homes have their windows facing the road rather than the sea is very telling. Newcomers to the coast are building houses with huge ocean-facing windows. It is rumoured that fisher-folks don't associate with "come from aways" because they fear *"catching what they got."*

life. Harvesters have several options: They can modify their lobster pots by using 100 per cent natural cotton twine to attach the entrance loops, mesh over a cut opening of the parlour, or attach the exit vents to the wire mesh with cotton twine. Many fishers complained that this conservation measure was too much work: the twine had to be replaced as soon as it broke down (every few weeks); the breakdown often wasn't noticed until a catch was lost. So they reached a compromise.

Harvesters now use iron clamps instead of stainless steel to keep an escape vent in place. In seawater, a 2-centimetre-long iron wire the size of a staple will rust away in a matter of months instead of weeks. This gives fishers a chance to get through the season without having to do constant repairs. Should a wire trap get lost at sea, the iron clamps holding the escape vent in place will rust away, causing the plastic vent to drop and give way for larger sea life to exit the trap. Measures to prevent ghost traps (pots that keep on trapping) are mandatory and strictly enforced.

Anatomy of the Lobster Trap

Just three decades or so ago, when woodlots were plentiful and free for the taking, lobster traps, also called "pots," were handmade from spruce slats and boughs. Fresh spruce boughs are very flexible; in early traps, they were bent to form the three hoops (front, middle, rear) that are the hallmark of traditional half-circle traps. Thinner spruce limbs were used to make the loops for the trap's two entrances. Fishers' wives often took on the task of making the hoops and twine netting from home as they raised their children.

Today, very few lobster harvesters make their own traps. The cost of wood, tools, and the time involved in building

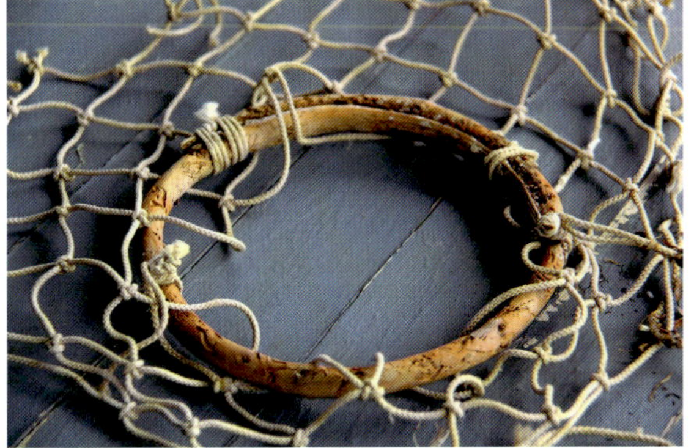

This turn-of-the-century trap loop was made from young spruce branches.

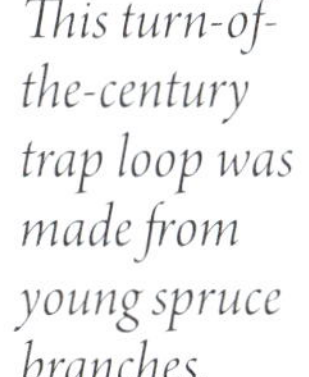

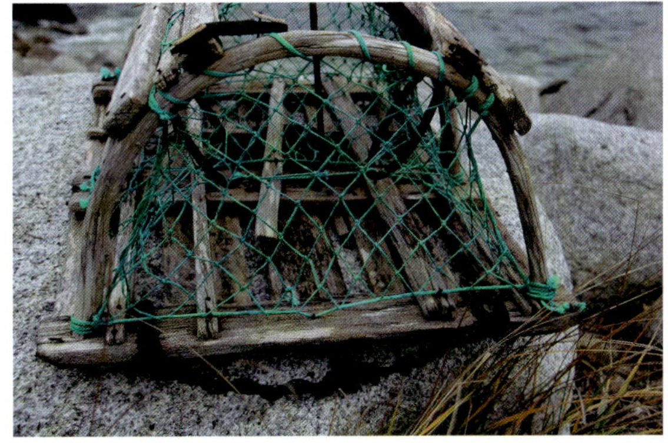

Few traditional half-circle traps made from spruce bough hoops are still in use. Those that remain often serve as ornamental garden accents.

Behind this fishing vessel are net-boats loaded down with gillnets. Mesh size is tailored to the desired fish species. Fish that are too large to swim through get snarled by their gills.

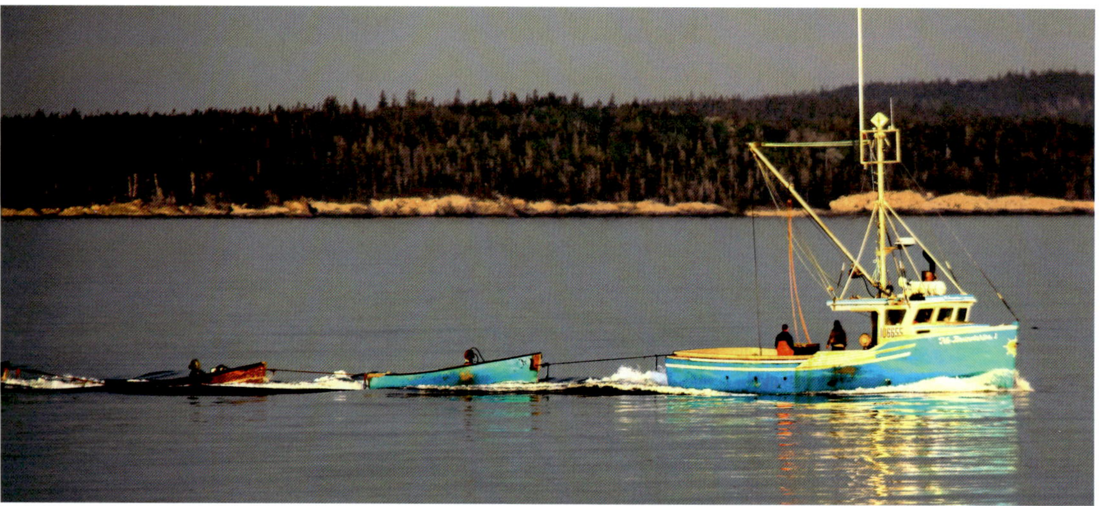

traps one by one is high. There are stringent specifications geared toward conservation that go into building a legal lobster trap. It is more efficient to buy traps built by specialized "trap builders." These traps and trap-building supplies are easily available all over the Atlantic provinces.

Modern, dome-shaped traps are made with durable engineered hardwood, plastic loop doors, and plastic mesh netting rather than natural twine. Although many lobster fishers still swear by them, wooden traps are being tossed aside for the more modern, rectangular, stackable, plastic-coated wire-mesh traps. Wood is susceptible to the wood-boring shipworm and becomes waterlogged. The square plastic-coated wire traps, on the other hand, weigh less and can be easily stacked, which allows fishers to pack more at one time on the boat. Another important feature of the lobster trap is ballast. These weights, usually composed of flat stones, bricks, or concrete runners, ensure the trap lands on the sea floor right side up. They also reduce a trap's chances of rolling around in strong currents.

Little Lobster Laugh

Where do French-speaking fishers get their supplies for building lobster traps?

Homard-ware (Home Hardware).

What Happens in the Trap *Doesn't* Stay in the Trap: Bait

A lobster trap is only as good as its bait. A big topic of conversation among harvesters is what others use as bait for landing such a good catch. Just about every harvester has experimented with unconventional bait, and some claim to have discovered their own secret formulas. Rumours fly that *this one* infuses the bait with molasses; *that one* dips his bait in bacon fat, and so on.... No matter the trick, the greatest challenge is keeping the bait in the trap long enough for the lobster to get at it.

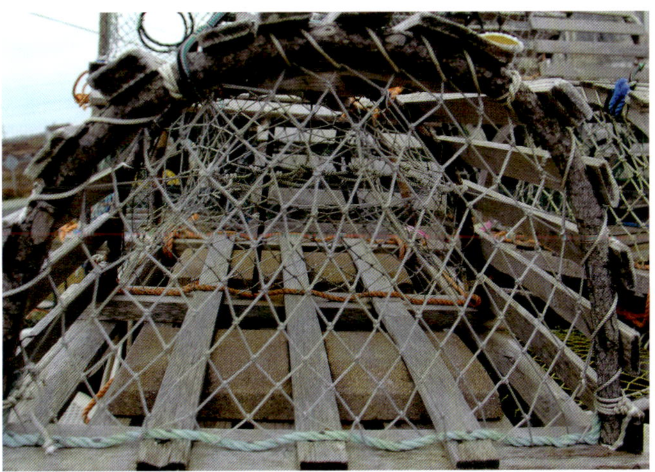

Ballast is built into the trap floor.

Not so long ago, a whole fish was simply forced over the sharp end of the large spike inside a trap. Some fisher-folk still do it this way, but voracious scavengers have the bait devoured within a few hours. More recently, bait has been placed in individual netted bait bags or wire boxes secured over the spike. But this still didn't keep small crabs and other marine creatures from getting at the meal before a lobster entered a trap. It was long thought that lobsters had

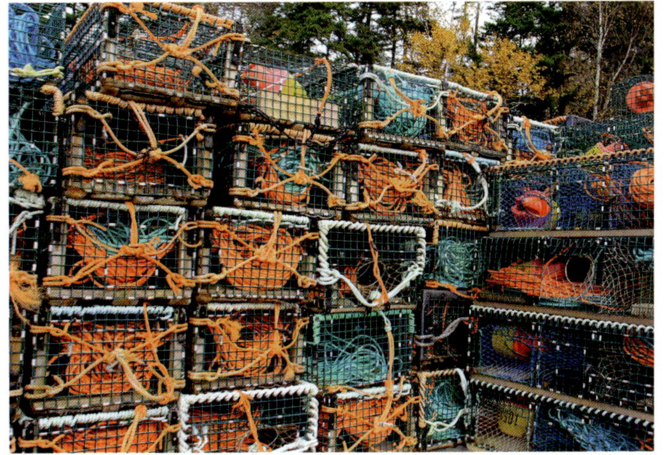

Modern, square wire traps are snugly stacked until the approach of Dumping Day.

LOBSTER TALE

The Lobster Fisher's Rug-hooking Wife

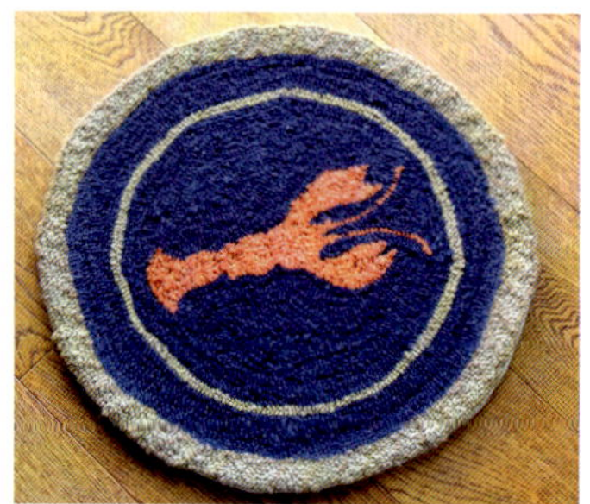

Hand-hooked rugs, a traditional Maritime craft, are cherished for their folk themes.

The doctor had just informed the ol' lobster fisher that his wife didn't have long to live, that he should call the family and get her affairs in order. The fisher returned to the hospital room to sit by his wife's side. He asked her tenderly if she had any last wishes. His wife told him he was to give their daughter what was in the box above the bedroom closet. When he got back home, the fisher immediately went looking for the box. Found! When he opened the box, inside were two rolled-up, hand-hooked doormats with a big red lobster in the centre. *Ah, Jenny will love this souvenir of her mother,* he thought.

When he lifted the mats out of the box, to his surprise was a pile of cash! The ol' fisher counted it up and to his astonishment, it amounted to almost $3,000! On his next visit to the hospital, he asked his wife what the two doormats were about. The wife answered, "Oh, every time I got mad at you, I'd hook one. The fisher gave this some thought and said, "Gosh, in our forty-four years of marriage you only got angry with me twice? That's pretty good." Then he asked her what the money was about. The wife answered, "That's from all the rugs I sold."

to *see* the bait in order to be enticed into a trap, but the search for a better bait bag led to the discovery that bait is nothing more than an olfactory attraction for lobsters, not a visual one. Lobsters actually have very poor vision. Furthermore, it isn't necessary for a lobster to actually have access to the food in order to be drawn in. A bait container that allows scent to be released but restricts access to scavengers is the way to go. It was hoped that the perforated neon-red plastic tubs would do the trick. But then there are the sea lice! Also called sea fleas, they are not much bigger than mosquito larvae. In swarms, they can devour the contents of a bait container within minutes.

Thus came the brilliant and most recent idea of the "flea bag," a sea flea-proof bait bag that slips over the bait spike. The durable, ultrafine nylon-mesh bag releases the scent of the bait inside while also keeping sea fleas out—completely. This has many

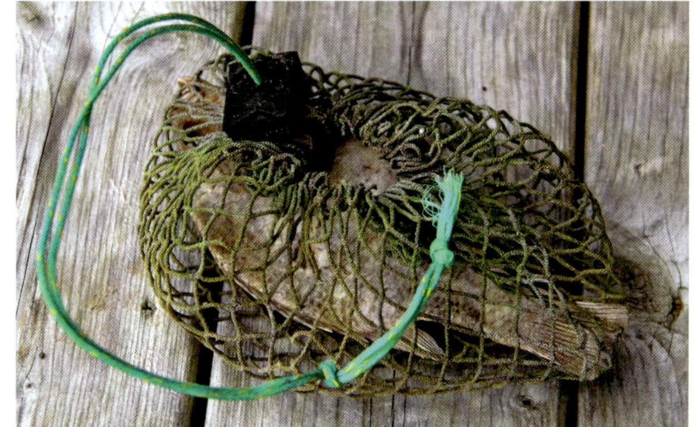

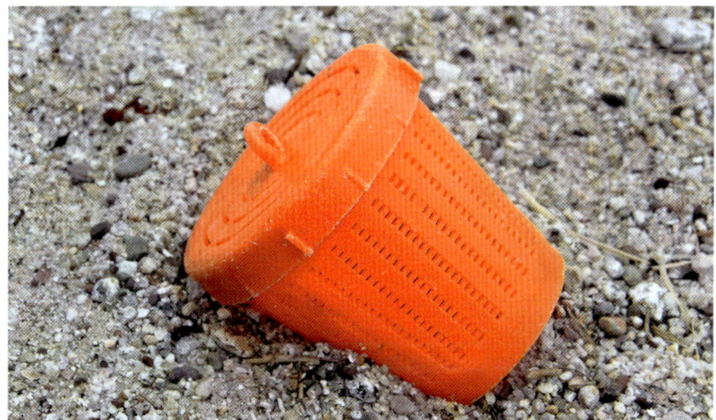

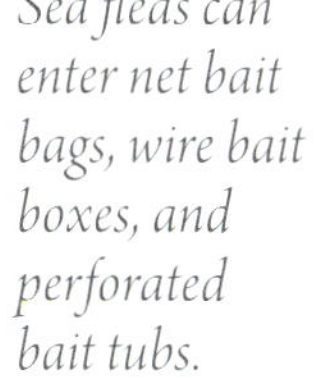

Sea fleas can enter net bait bags, wire bait boxes, and perforated bait tubs.

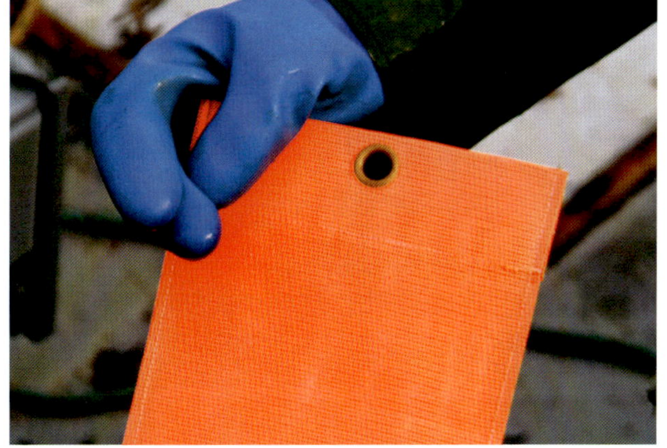

A sea lice–proof bait bag.

benefits: The bait will last longer, so less bait is wasted. Since there is less need to fish for bait, fewer marine species are harvested for that wasteful purpose. With less access to bait, sea lice populations are reduced. Furthermore, since the bait will last longer, traps can be left in the water for longer periods. Lobster fishers are no longer as pressured to check their traps in inclement weather.

The Charms of Hardworking, Handmade Buoys

The fact that buoys are visually appealing as a result of their functionality is quite rare in the world of design. Their tapered shapes (to prevent drag when pulled out of the water), sizes, and colours are purely practical, their beauty completely accidental.

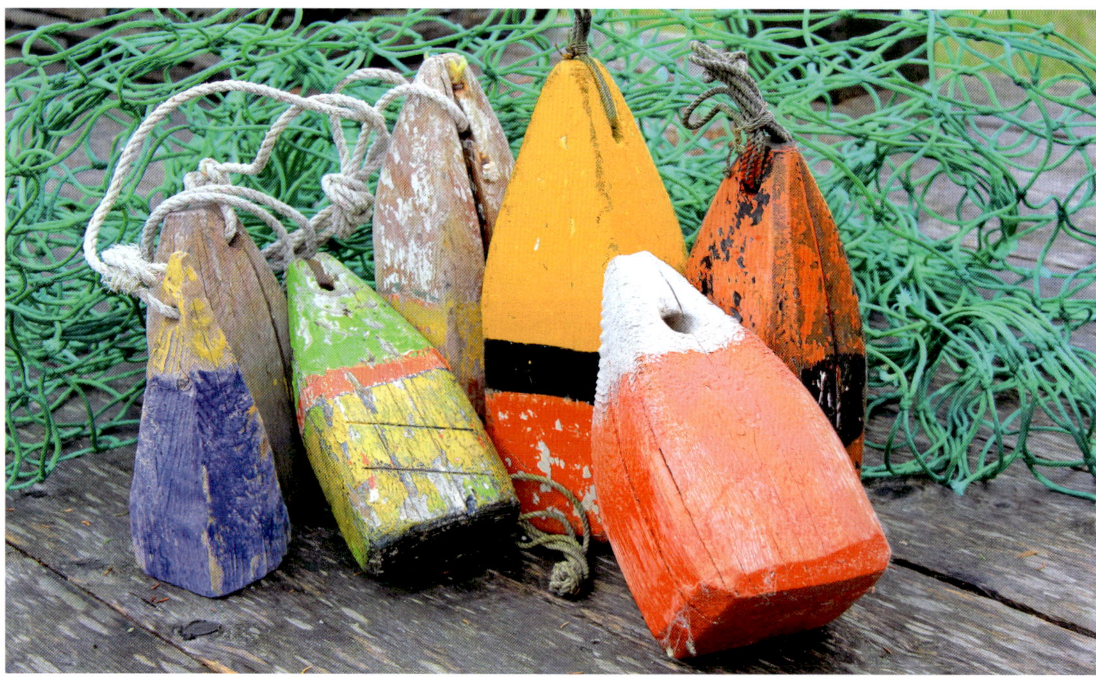

Buoys from the good ol' days, when wood was plentiful and free.

"Some stuff freaked me out when I first began lobster fishing with my boyfriend. I remember when we pulled up a trap that had been baited with a whole fish on the spike. I opened the trap lid and saw a fish wiggling vigorously. I thought: *Is it possible that's alive? How can that be? We don't use live fish for bait.* When I pulled the fish off the spike, sea lice by the thousands spewed out! There were so many sea lice inside the dead fish's cavity that they made it move like it was alive! Most of the time though, all there is left when we pull the traps up is a fish skull and bones." –**Lynn Natte, former lobster fisher, Peggys Cove, Nova Scotia**

As with wooden traps, classic painted wood buoys, too, were originally handmade from trees native to the Atlantic Canadian coast. The best and most desirable type of wood was white cedar. Its fibres contain an oil that naturally repels water so cedar floats were much more buoyant and lasted longer than those made from other native trees. As a result, cedar was less prone to getting waterlogged or decaying, and hardly ever showed signs of aquatic insect infestations.

But cedar wasn't readily available to all fishing communities. The ever-present spruce was the most commonly used wood, especially in Nova Scotia and Prince Edward Island. Early on, wooden buoys were dipped in tar diluted with gasoline for waterproofing and longevity. For environmental reasons, this practice was no longer permitted by the 1960s and '70s. Instead, painting wooden buoys became a more common method of preservation, and it had the added benefit of allowing each harvester his or her own colour scheme for trap identification. The fishers chose their buoy colours on an honour system. The colour schemes of those who fished the same waters had to look markedly different. This custom of colour-coding buoys is still practiced to this day, though the materials buoys are made of have changed.

The floating marker of choice today is made of Styrofoam, as it is much lighter and more buoyant than wood and does not get waterlogged. As such, authentic, well-worn wooden

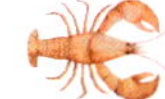

buoys are becoming increasingly rare. Handmade and hand-painted wooden buoys may only be found in local craft and antique shops. Avid collectors seek out antique dealers in search of well-preserved, retired wooden buoys, and many are willing to pay a premium for a fine specimen. These wooden hand-painted buoys are prized not only for their aesthetic allure but because they are true relics of Atlantic Canada's fishing heritage.

While wooden buoys are no longer made for fishing purposes, a few artisans make them for tourists. And none is as passionate about making buoys as Paul Gallant, a semi-retired accountant from New Brunswick. Gallant not only makes wooden buoys, he collects them. He knows everything that could possibly be known about the history of buoys in the Maritimes. His property in Grand Barachois is a veritable museum of colourful buoys from generations passed and present.

To Gallant, buoys that had a working life at sea are the most interesting: they have a story to tell. Their shape, size, colour, and state of deterioration hold important clues. From

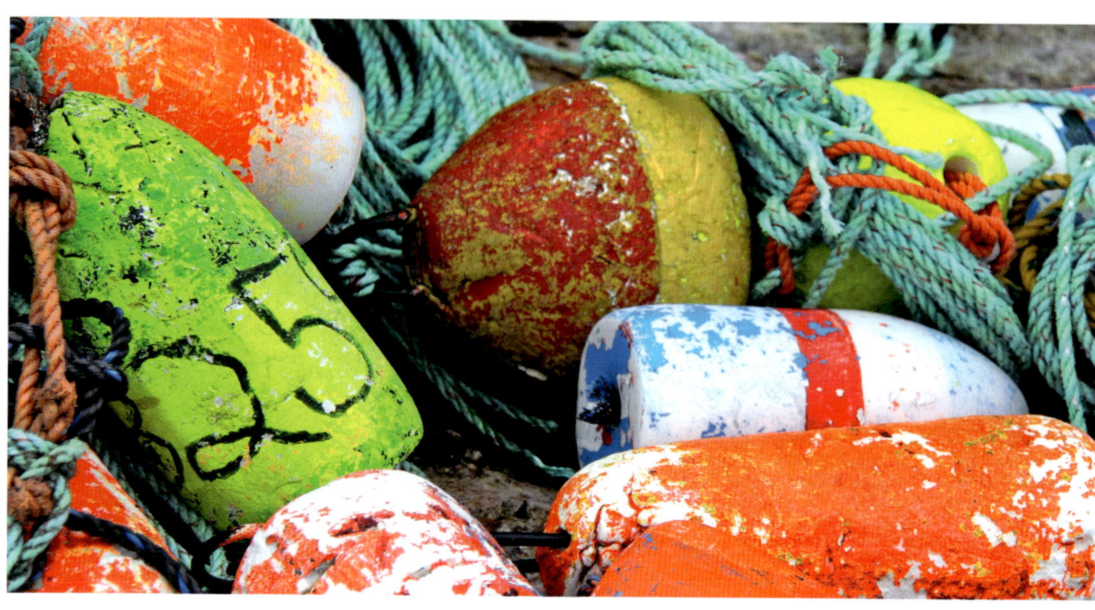

Whether made of wood or another synthetic material, buoys are always tapered at one end where the rope is attached. This creates less drag when pulling a buoy out of water.

these, Gallant can determine where the buoys are from, when they where in use, and what their specific function would have been. Some were used for the inshore fishery, others for the offshore fishery. (Traps linked together by a few metres of rope required a different buoy design than those marking a single trap.)

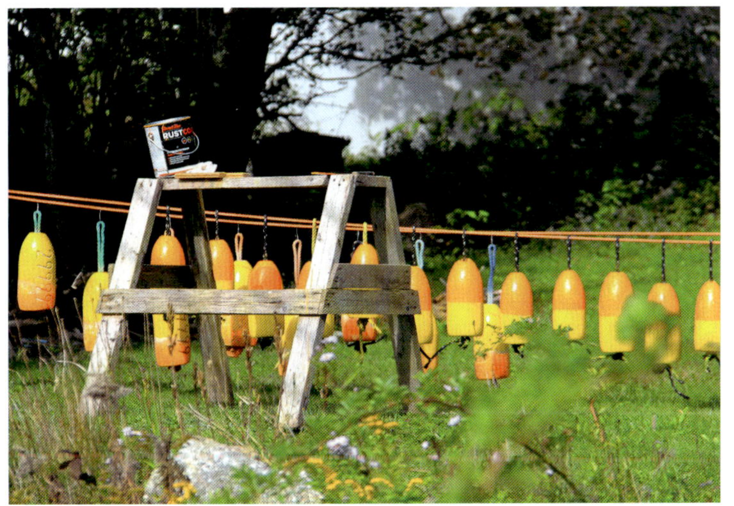

Freshly painted Styrofoam floats in yellow and orange hang to dry on a line just before the fishery opens.

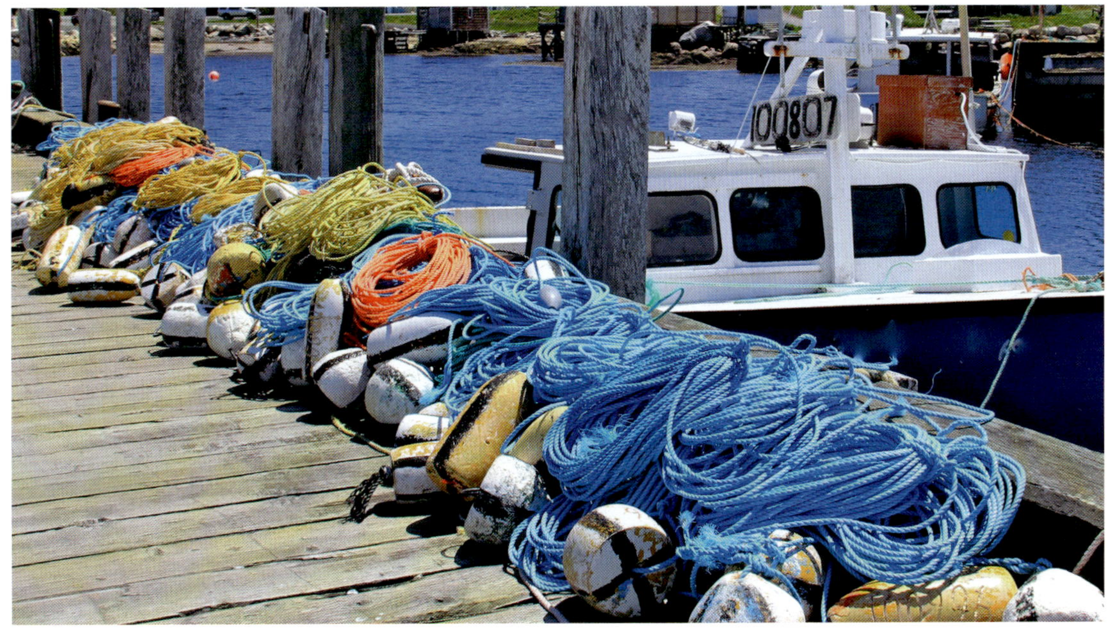

There is no shortage of colour at this wharf next to Ryers lobster pound in Indian Harbour, Nova Scotia—and every day is different.

Paul Gallant's buoy collection displayed at the shop where he works. Some are nostalgic buoys he makes for himself; others are made especially for selling to tourists.

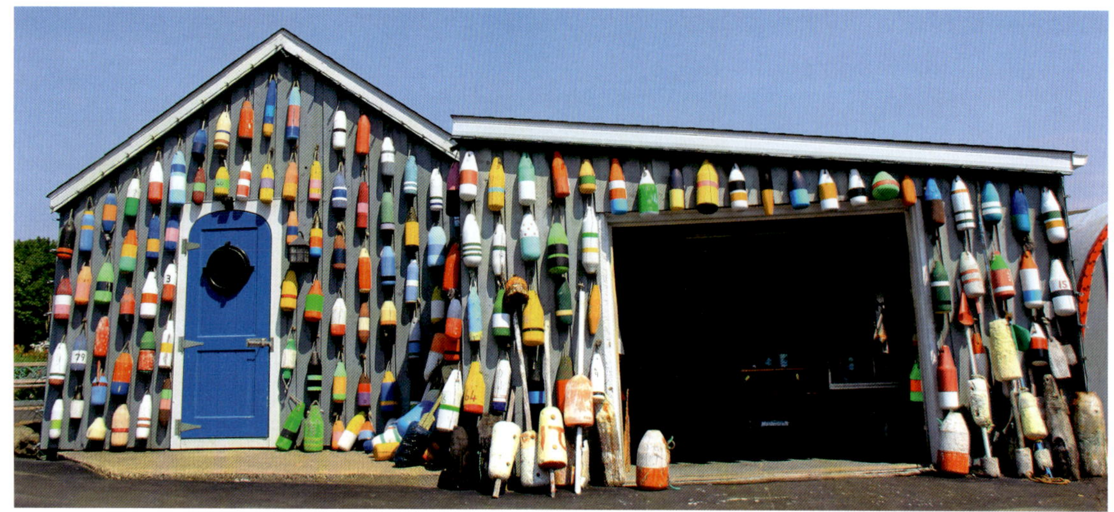

Gallant has also perfected the craft of buoy-making with white cedar, as it was done in his father's day:

> *My father, Clairice Gallant, was a fisherman all his life. He died at age ninety. I fished lobster with him at one time, for several years. He supported our family of seven kids by fishing lobster in the summer and smelt in the winter. It was a hard, backbreaking way to earn a living. In those days, there was no technology like GPS [and] depth sounders, and there were no motorized winches for lifting the traps like they have now. Those wooden traps, when wet, can weigh up to a hundred pounds! It was all done by hand. Where to set the traps was all guesswork, experience, and luck. Setting nets for smelt under sea ice was no easier and just as risky as fishing for lobster out of an old wooden boat way out on the water.*
>
> *I showcase these buoys because they were so important to fishing. They are a symbol of resilience and hard work. I love them. I think they are beautiful. They bring back wonderful memories. I collect them and make them in honour of my father's life at sea.*

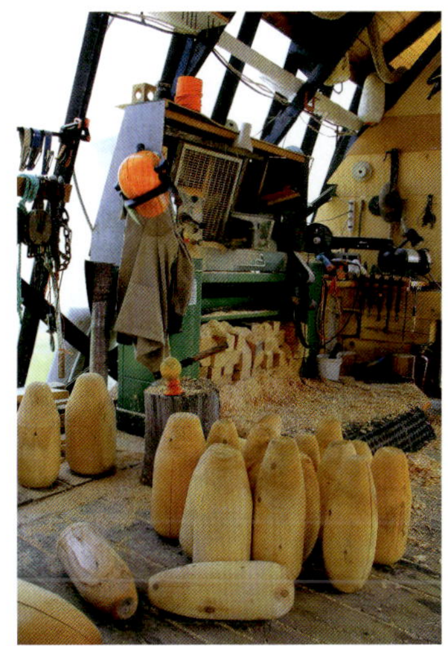

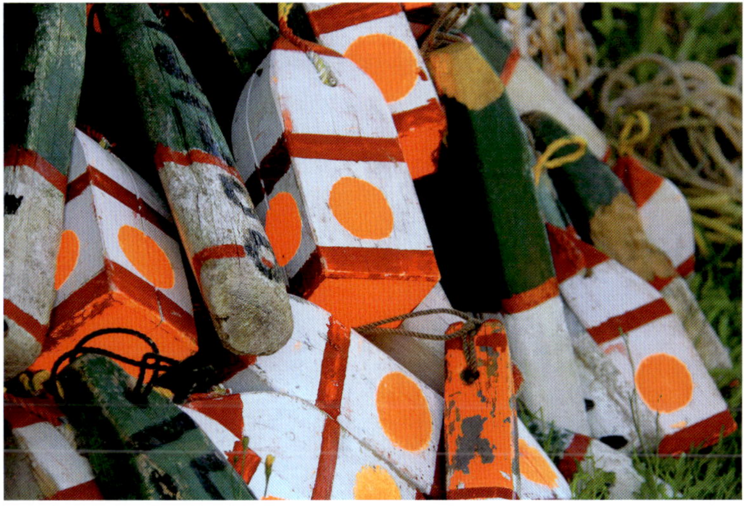

(Left) Paul Gallant's cedar buoy workshop. (Right) Trap ownership is identified by buoy shape and colour pattern.

Gallant begins in the fall with "green" wood (fresh-cut cedar). Working with the still-wet soft wood means less wear and tear on his tools. He then leaves the beautifully lathed buoys to cure in his shop over the winter months. Come the warmer days of spring, his wife, Noella, paints the buoys in colours and patterns that local fishers still use to this day. The array of colours make an irresistable display. Paul Gallant sells his buoys roadside from his home, mostly during the summer months. There is no need to advertise. Enthusiasts come upon the attractive, shiny new buoys either by word of mouth or by chance.

LOBSTERING PART II: *a* ROUGH *and* RISKY OCCUPATION

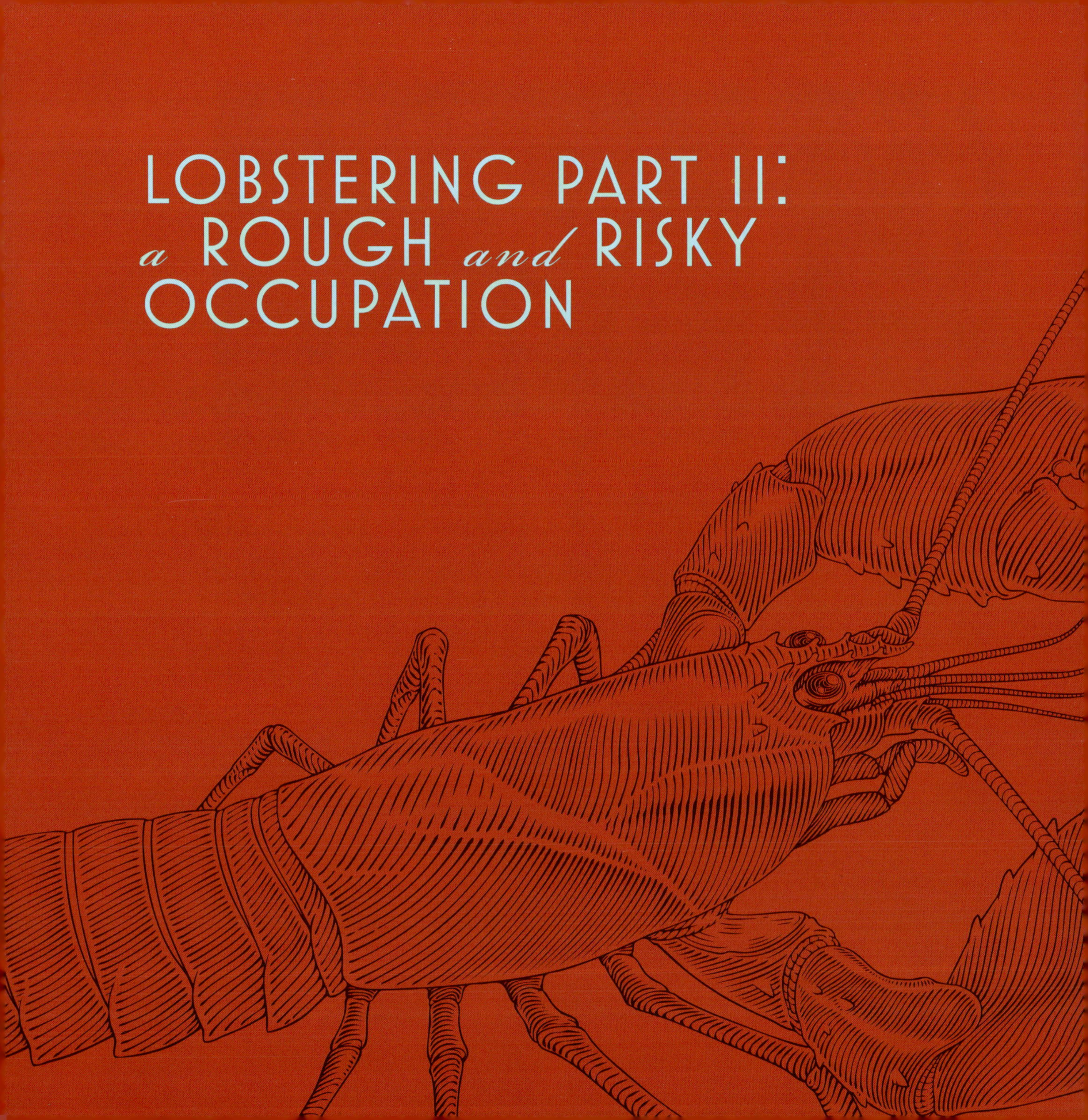

Blessing of the Fleet

We often, and perhaps naively, stand guilty of romanticizing the lifestyle of those who fish for a living. "Lobstering" is not easy work. In fact, earning a livelihood at sea is one of the most dangerous occupations in Atlantic Canada. It should be no wonder that many fishing communities hold an annual "Blessing of the Fleet," a heartwarming faith-based event where loved ones and friends gather to wish the fishers a safe season. Most fishing villages have one or more churches. They are often home to works of art depicting the hardships of fishing and memorials to those who were lost at sea. Many epitaphs in fishing-community cemeteries show symbols of a life won or lost on the ocean: a boat, an anchor, a compass, a ship's wheel....

A popular and well-photographed spot in the fishing village of Peggys Cove, Nova Scotia, is the summer home of the late, celebrated marine artist William E. deGarthe (1907–83). Born in Finland, deGarthe immigrated to Canada in 1926 and taught at the Nova Scotia College of Art in Halifax (now called the Nova Scotia College of Art and Design University). Every spring he would stay at his house in Peggys Cove, where he painted scenes of surrounding seascapes and

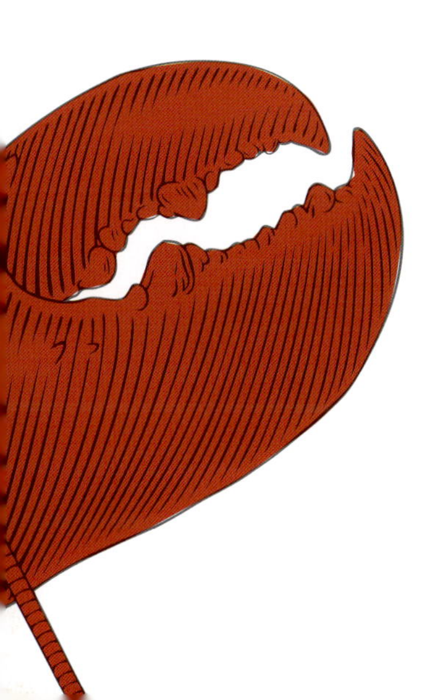

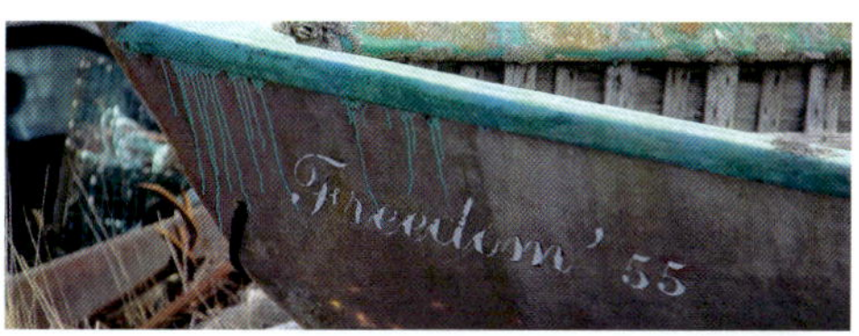

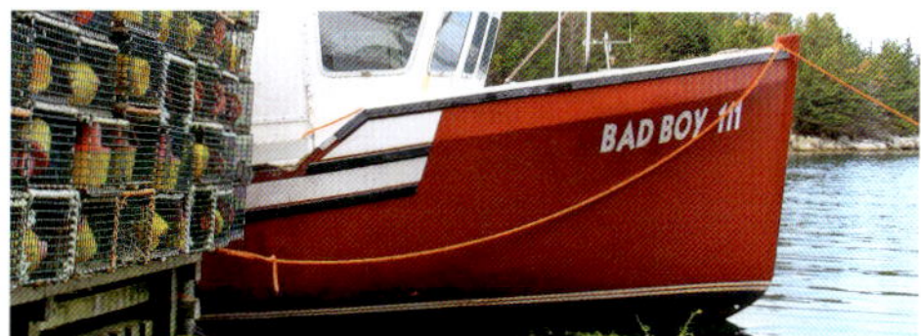

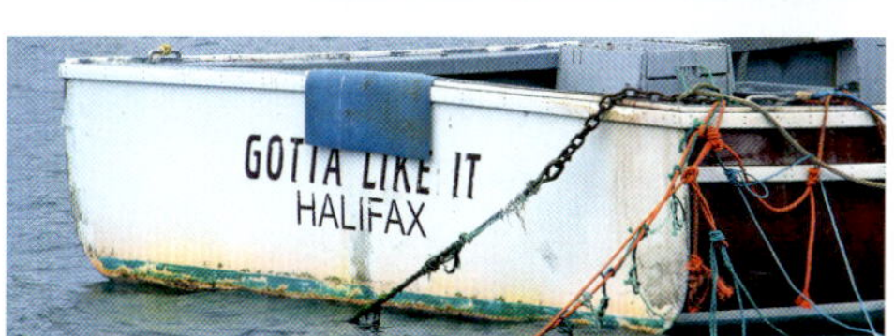

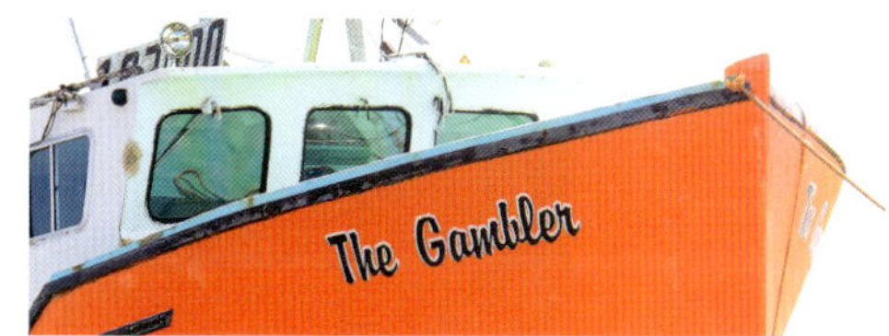

Boat names reveal much about attitudes toward working at sea. The spectrum is wide—from respect and veneration to a bit of flippant humour.

William E. deGarthe (1907–83)

Born on a remote Finnish island town known as Kasco, William E. deGarthe showed an early aptitude for the visual arts. Shortly after high school, he was called for active duty in the Finnish military. After his release from service, deGarthe obtained a passport, declared his profession as an artist, and immigrated to Canada in the fall of 1926. He first landed in Halifax and from there boarded a train to Toronto to join other Scandinavian expatriates who were employed in the forestry trade. The hard working conditions and biting winters caused the penniless nineteen-year-old to flee to Montreal, where he found his way to a mission. During his stay, young deGarthe showed a few of his drawings to the supervisor. This led to his being introduced to a local publisher, who hired him as an illustrator for seven dollars a week. The young deGarthe was able to attend formal art studies at the Montreal Museum of Fine Art.

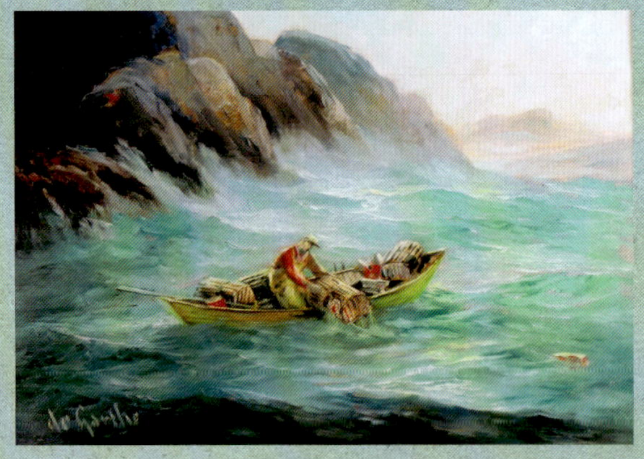

Hauling Lobster Traps *by William E. deGarthe, 1971. Oil on canvas. (Nova Scotia Archives)*

In 1930 deGarthe's thirst to find the most beautiful place on earth to paint led him to quit his job as an illustrator and get on a train. He pursued art studies in various parts of Atlantic Canada, the United States, and Europe. In 1955 he set foot on Nova Scotian soil once again and, struck by its similarity to the coast of his native Finland, he declared: "I don't have to travel any farther."

Today deGarthe's paintings can be found in homes and offices all over the world, but his most memorable and valuable artworks are those that depict people who make their living at sea in the area of Peggys Cove, Nova Scotia.

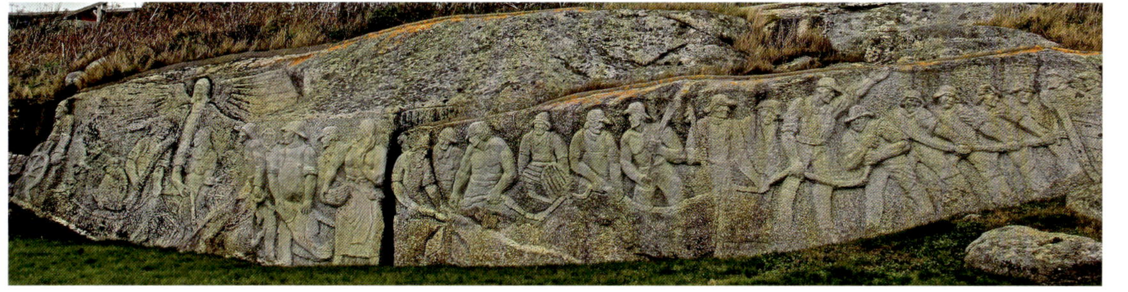

William deGarthe's moving tribute to fishermen and their families, Peggys Cove.

resident fishers at work in their boats. In 1977 deGarthe donated a collection of paintings to the Province of Nova Scotia that spans forty years of his life as an artist; he wanted his work preserved as a historical record, stating, "Over the years, I have purposely held back on selling these paintings that portray an era in the province's history that no longer exists."

Most people learn about deGarthe when visiting Peggys Cove, where stands his formidable stone carving out of a one-hundred-foot-long natural granite ledge in his former yard. It depicts a guardian angel standing over the community of fisher folks. He began the relief sculpture in 1970 and continued to work on it for the next ten years. It was, to him, only half-finished by the time of his death. His charming old house is an important landmark that now serves as a museum showcasing his marine paintings and sculptures.

The Hazards Of "Dumping Day"

The momentum is building. You know Dumping Day is fast approaching when the sound of hammers, drills, saws, and squealing winches accompany the call of seagulls. Your boat better be seaworthy and ready to take the pounding waves. Your ropes and traps must be in the best of shape for the hard job ahead.

"Dumping Day" is a common term used in fishing communities for the opening day of lobster season. The Atlantic lobster fishery is divided into forty-one sections. Lobster fishing

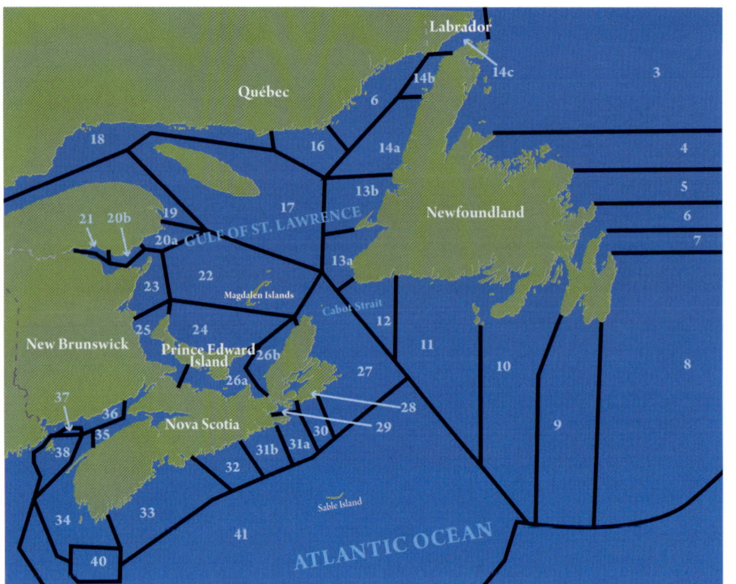

Fishing districts of the Atlantic provinces, determined by the Department of Fisheries and Oceans.

season takes place at different times in these zones throughout the Atlantic provinces. Staggered seasons over the course of a year ensure a steady supply of lobster to markets year round. Some communities, such as those in Prince Edward Island and New Brunswick, have a short summer season, while others, like the South Shore of Nova Scotia, districts 34 and 33, have a long winter season that goes from December 1 to May 31. Nova Scotia has sixteen districts (25–41), a few of which are shared with New Brunswick and PEI. Dumping Day is a momentous day no matter what time of year lobster season opens. All at once, at 7:01 A.M. sharp, the harvesters head out to "dump" their baited traps over the side of their boats.

Everyone, it seems, has a story to tell about Dumping Day. It's even the title of a popular song by Ed Saunders, Terry Saunders, and Brian Smeltzer of *The Saunders Brothers Show*. The group was inspired to write the upbeat song because Dumping Day is such a highly anticipated event in Peggys Cove, where they live. The lyrics are very telling of the excitement, the hard work involved, the camaraderie—and what can go wrong. It reads like a poem. The place names, individuals, and family names mentioned are real.

"DUMPING DAY"

The Saunders Brothers Show

Good Morning, Mr. Manuel, Graham says the coffee is on
Get in your gear, there's bait to break up, and we're pulling out before dawn
It's dark when you leave and it's dark coming back
Passed Dover Castle and Halibut Rock
And the fish cop is waiting up in his truck
With spy glasses on the Bay
Hey! It must be Dumping Day!

Hello there, Mr. Morash, we're steaming up past Paddy's Head
Your gloves and your pants are all dried off, hanging upstairs in the shed
With traps piled high and coiled buoy lines
Wait for the signal then over the side
It was calling for calm winds; we left with the tide
But now it's blowing a gale!
Hey! It must be Dumping Day!

The Hubleys, Fralicks, McRaes, and Ryers
Neighbours and drinkers and lovers and liars
It's round the breakwater, past Peggy's Point
Head in the west and away....

Good day to you there, Mr. Schnare. Have you got a joke to tell?
The venders are lining up for a price—roadside canners to sell
The radio crackles: a boat has broke down
Toss them a line and then haul them around
We are in this together, brother to brother
Shove off with no delay! It must be Dumping Day!
Hey! It must be Dumping Day!

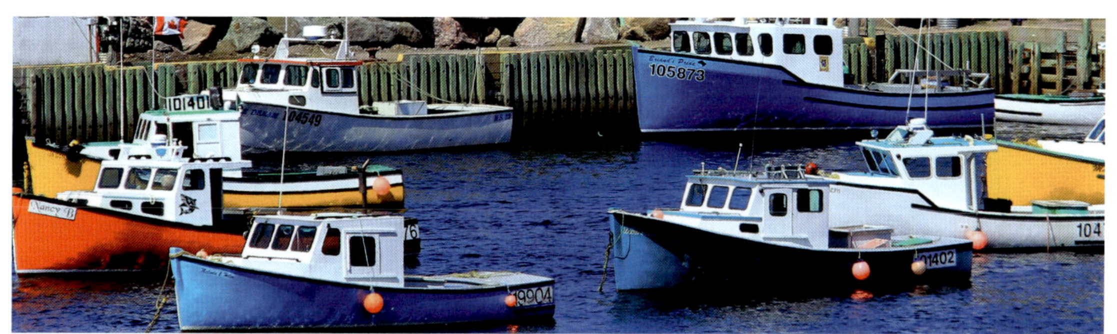

A fleet of lobster-fishing boats lay idle, awaiting the opening of the fishing season.

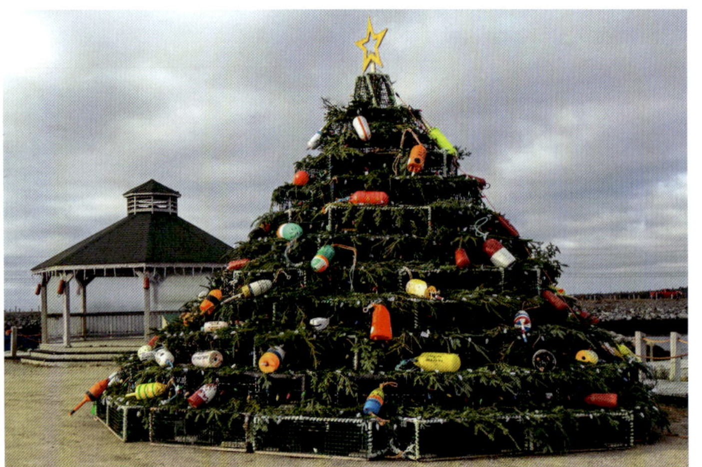

A fishing community shows appreciation for the hard-working lobster harvesters.

The first day of the lobster season is the most important but also the most dangerous. The primary concern is weather. Even in the best of conditions, it wouldn't take much of a swell to tip a boat weighed down with traps. It is not unusual for Dumping Day to be postponed due to high winds and rough seas, especially for those districts with a late fall to early spring season. Fall is hurricane season in Atlantic Canada, after all. Lobster harvesters don't have much patience for weather-related delays. Every day lost is income lost. If the forecast is severe, Dumping Day is delayed for all involved. When the forecast calls for varying conditions, it is up to each fisher's discretion to stay in port or to go out. Because the first few days of the season bring the best catch and income, the majority will choose to take their chances out on the water. Harvesters bring in more than half the entire season's catch of lobsters in the first few days following Dumping Day.

Needless to say, the stakes are high. And lobster harvesters are fiercely competitive. Each fishing community has "traditional" fishing grounds. (Peggys Cove fishers set their traps in waters near Peggys Cove. Boutiliers Point fishers set theirs in waters near Boutiliers Point… and so on.) On Dumping Day, everyone races out to what are perceived to be the best spots in the dark hours of the morning, before someone else takes them over. Traps are set between 50 and 100 feet apart all over the ocean floor. Fishers with a Class A license can set a maximum of 250 traps, while Class B license holders have a 75 trap limit.

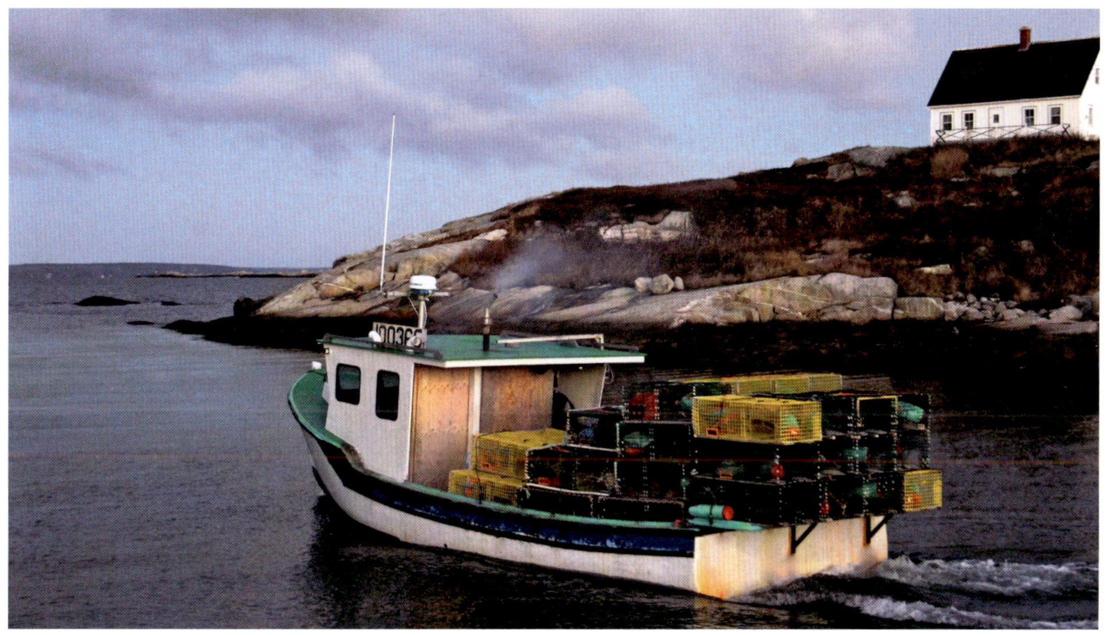

One by one, loaded with traps as high as they can be stacked, boats head out of the small picturesque harbour of Peggys Cove, Nova Scotia.

The boats all head out at once at a time determined by DFO—usually 7:01 A.M. sharp—and a coast guard ship is sent out to patrol the waters. Time is of the essence for lobster fishers: they need to get all their traps set as quickly as possible, making several return trips to the wharf to load up more traps until every one is set out to sea. Their boats are loaded to capacity, with traps weighing 80–100 pounds each.

Stacking the traps is an art in itself: piled up too high, they could cause the boat to take on water. A rogue wave could cause the traps to spill overboard and hit an unsuspecting worker. When unloading the traps, keeping the weight balanced between port and starboard is very important: the boat could tip over if one side is heavier.

Harvesters can only hope that each trap dumped will land where there are hungry market-size lobsters nearby. The appearance of so much free food (bait) all of a sudden and all at once is pretty exiting to a creature used to having to search long distances for a meal.

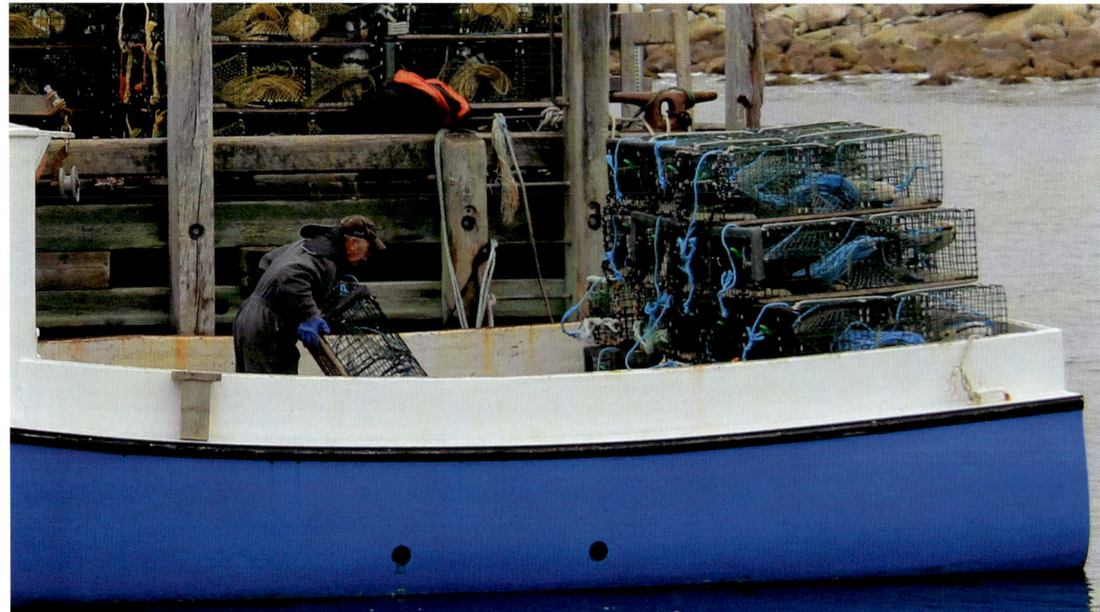

Jeff Boutilier of Indian Harbour, Nova Scotia, makes his third trip back to the wharf to load more traps.

The traps keep coming down throughout the entire day. Some fishers will still be at it well into the evening, depending on how far out from shore their traps are set. There is no rest until the last trap is dumped, the last attached buoy thrown to sea. This is a moment of great relief—but the joy of rest is short lived.

Dumping Day is an exhausting day. But it's not over yet.

Lobsters are most active at night. Once all the traps have been dropped to sea, lobster fishers head back home for a short break. At one minute after midnight, they are permitted to begin checking their traps. Those with sleeping quarters on board often stay at sea. At twelve o'clock sharp, powerful lights are turned on and the second race is on. Already fatigued from a hard day's work, most lobster fishers will go all through the night again, checking and baiting their traps one at a time for lucrative lobster. "There's barely time to wipe y'er nose in this business" said one fisherman, "and a lot can go wrong." This is gruelling, backbreaking,

"I went fishing for lobster for fifty years. I started quite young 'cause I didn't like school much. Back in them days you could go fishing all you wanted at any time of the year. We didn't bother with giving the boat a name. We was just workin' for a livin.' There was no taggin' traps and that and all the rules they have today. I made my own traps from spruce.

Anybody could get a lobster license for a quarter. I sold my lobster license for $100,000 just a few years ago. Today a lobster license goes for $250,000, easy. The fellers who fish around here got into it because it was passed on from the grandfather, to the father, and then the son. There's no other way to get into it."
–Edward Richardson, lobster fisher, Indian Harbour, Nova Scotia

repetitious work. Next time you sit to dine on a succulent freshly caught red lobster, be mindful of the hard work and risks that went into catching your meal—and don't dare complain about the price!

Thousands of pounds of lobster per boat are landed on those first few nights. Harvesters are exhausted and sore. The crankiness soon fades when they see what they've made in that first week of the season—for many, the income equals an average family's annual salary.

The high presence of the Canadian Coast Guard on this first day is meant not only for law enforcement but also to ensure the safety of the lobster harvesters. Air rescue squadrons remain on standby in case someone gets in trouble. Fires and mechanical malfunctions are all too

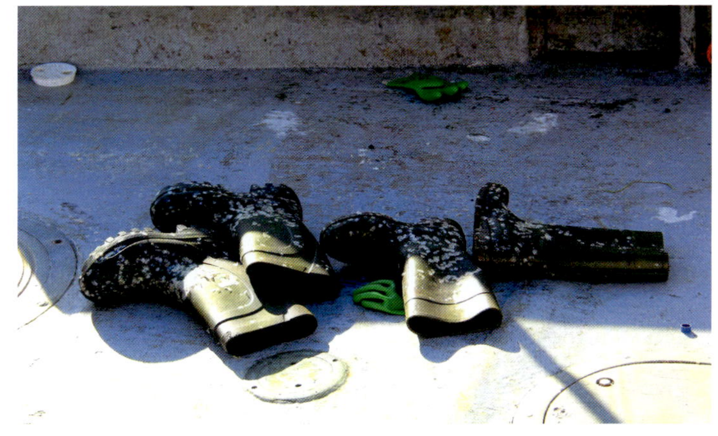

A sign of fatigue at the end of a long day at sea.

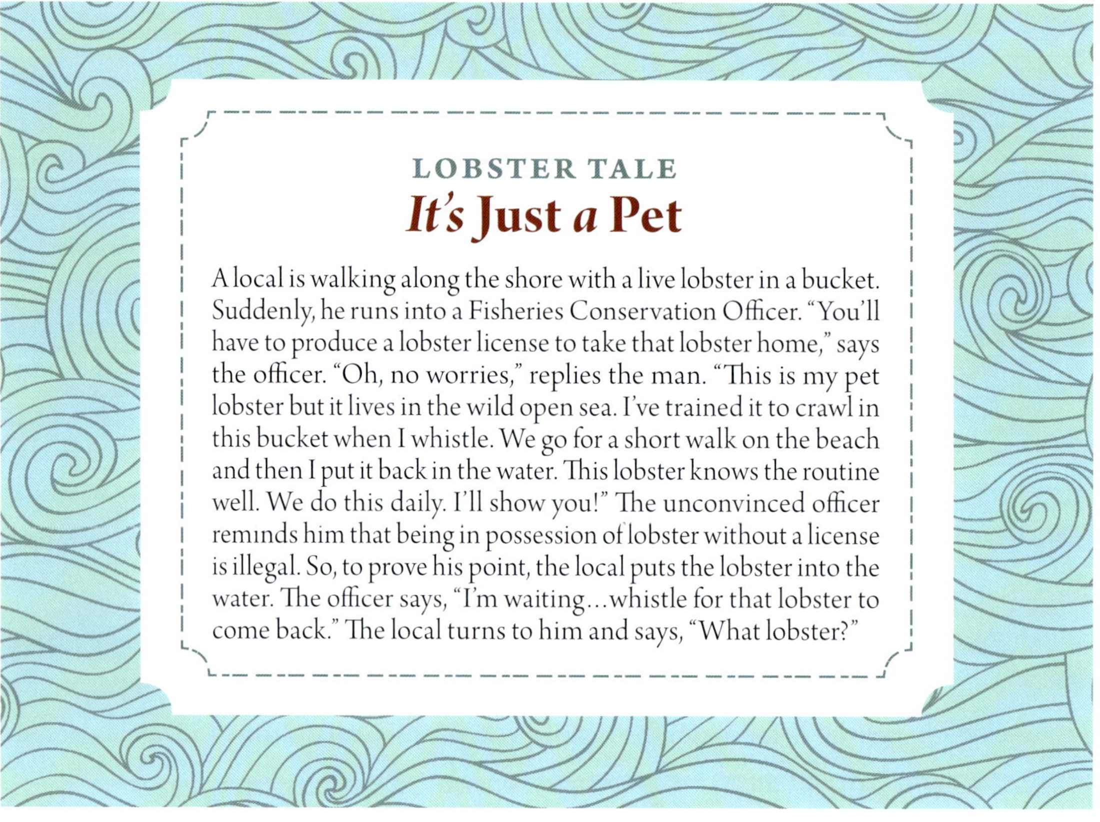

LOBSTER TALE

It's Just *a* Pet

A local is walking along the shore with a live lobster in a bucket. Suddenly, he runs into a Fisheries Conservation Officer. "You'll have to produce a lobster license to take that lobster home," says the officer. "Oh, no worries," replies the man. "This is my pet lobster but it lives in the wild open sea. I've trained it to crawl in this bucket when I whistle. We go for a short walk on the beach and then I put it back in the water. This lobster knows the routine well. We do this daily. I'll show you!" The unconvinced officer reminds him that being in possession of lobster without a license is illegal. So, to prove his point, the local puts the lobster into the water. The officer says, "I'm waiting...whistle for that lobster to come back." The local turns to him and says, "What lobster?"

common with the mad rush out to sea. A sinking boat, a man overboard, a heart attack, these seem to happen every year on Dumping Day somewhere on the Atlantic—and unfortunately, so does the occasional drowning.

Mandatory safety gear varies depending on the size of the boat and whether it is an inshore or offshore vessel. Inshore fishery boats, which are smaller, can range from a 13-foot skiff to a 30-plus-foot Cape Islander. Offshore boats, typically called trawlers, can range from 50 to 90 feet in length.

The Good Ol' Seaworthy Cape Islander

A Cape Islander rests on a skid above the high-tide mark awaiting a paint job.

The so-called Cape Islander–style fishing boat is a Nova Scotia icon. The most commonly used vessel of Atlantic Canada's inshore fishery, it is famous for its long stern for piling traps and nets, and its wide, upward-slanting bow for ease in riding large swells. Two families claim credit to the invention of the Cape Islander between 1905 and 1907. Both are from the town of Clarks Harbour on Cape Sable Island, Nova Scotia, after which the boat design is named.

Personal floatation devices (PFDs) must be on-board all boats but are almost never worn because they restrict mobility, making it too cumbersome for fishers to get around safely. Because of the hard physical work involved in lobster fishing, PFDs also cause uncomfortable heavy sweating. Larger fishing vessels and offshore trawlers with more staff must have survival suits on board as well as a life raft, a tool box, a fire extinguisher, flairs, a first-aid kit, and a VHF (very high frequency) communication device. Many smaller, inshore boats carry nothing more than a lifesaver ring and a simple life jacket for each person aboard. These brave harvesters will tell you all that fancy safety gear is "just there for looks"; that the frigid temperature of the ocean will kill you before you get a chance to get in a suit or a lifeboat. The way fishers see it, putting one's life on the line is just part of earning a livelihood at sea.

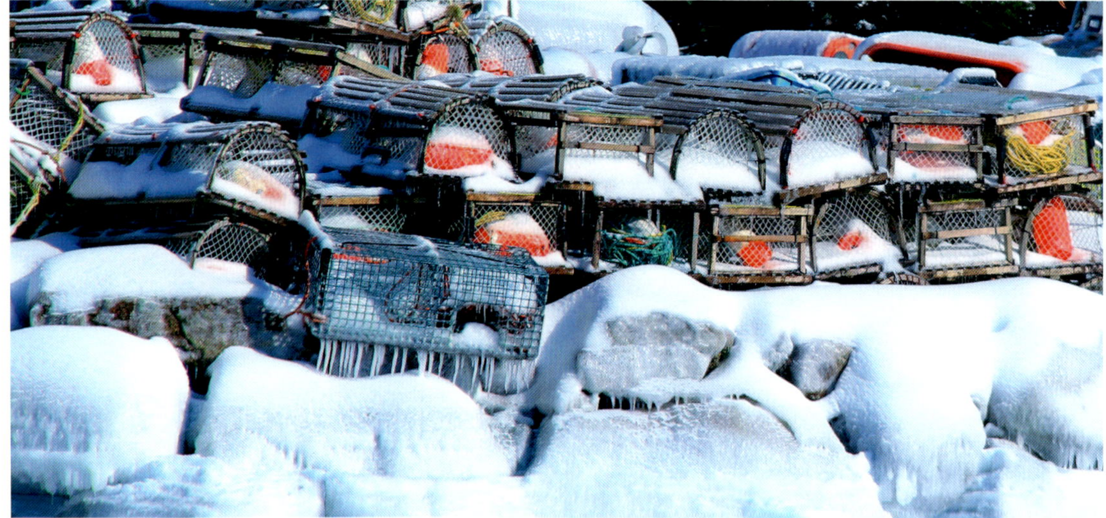

Traps lay idle on a frigid February day in Modesty Cove, Indian Harbour, Nova Scotia.

Lobstering Dangers

WEATHER

Once the season is open and the lobster fishery is in full swing, it is up to each individual fisher to schedule when and how often to head out to sea. There are only two things that dictate how much energy to put into lobster fishing: the weather, and how good the catch is. For those areas with a late fall to spring season (a winter fishery), lobstering will continue only until it becomes intolerably cold. Frigid water causes lobster's metabolism to slow down and they become inactive; their need to eat and appetite drop no matter how tantalizing the bait. When the cost of fuel exceeds the return on low catches, it's no longer profitable to keep fishing. Tired and weary lobster fishers pull their traps up and take a deep-winter break at this time, usually from the last week of January to mid-March.

However, with climate change, this trend too is changing. The frequency and warmth of winter thaws seem to be on the rise. In the last few years, Tim Hubley of Indian Point, Nova

"When I was goin' out, we never had no bad accidents when fishing for lobster. We just watched the weather. Today, they put too much trust in the high-tech gear and take too many chances. That's why some of them get in trouble."
–**Edward Richardson, lobster fisher, Indian Harbour, Nova Scotia**

Scotia, has been venturing out to check his traps most days throughout the winter months. He believes Nova Scotia winter waters are indeed warming and that translates into more lobster activity. He also feels there may be more lobsters than there have been historically in these parts. It's only speculation at this point, but Nova Scotia may be reaping the benefits of Maine lobsters moving north.

Come early spring, after months without income, lobster harvesters are anxious to return their traps to sea. This time of year has its risks. Some inner harbours get iced over. To ensure easy passage, boats must go in and out of their coves periodically in order to keep the ice broken while it is still relatively thin. Keeping the ice fragmented is imperative to making it out of coves to the open sea. But what to do if ice forms around a moored boat and the only way to reach the vessel is by rowboat? Well...you take your chances.

Ice was the cause of a tragedy at sea on January 6, 2014, in Lunenburg County. The body of a seventy-nine-year-old man reported missing was found submerged near his boat. It appears he was trying to break the ice from around it and slipped out of his skiff into the frigid water.

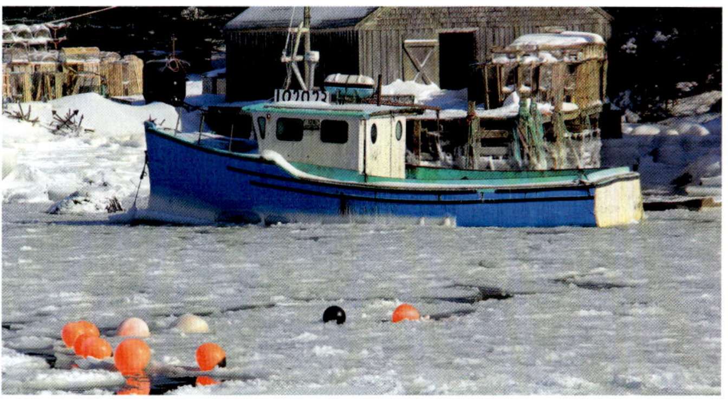

The deep freeze has set in. Time for a rest until the spring thaw.

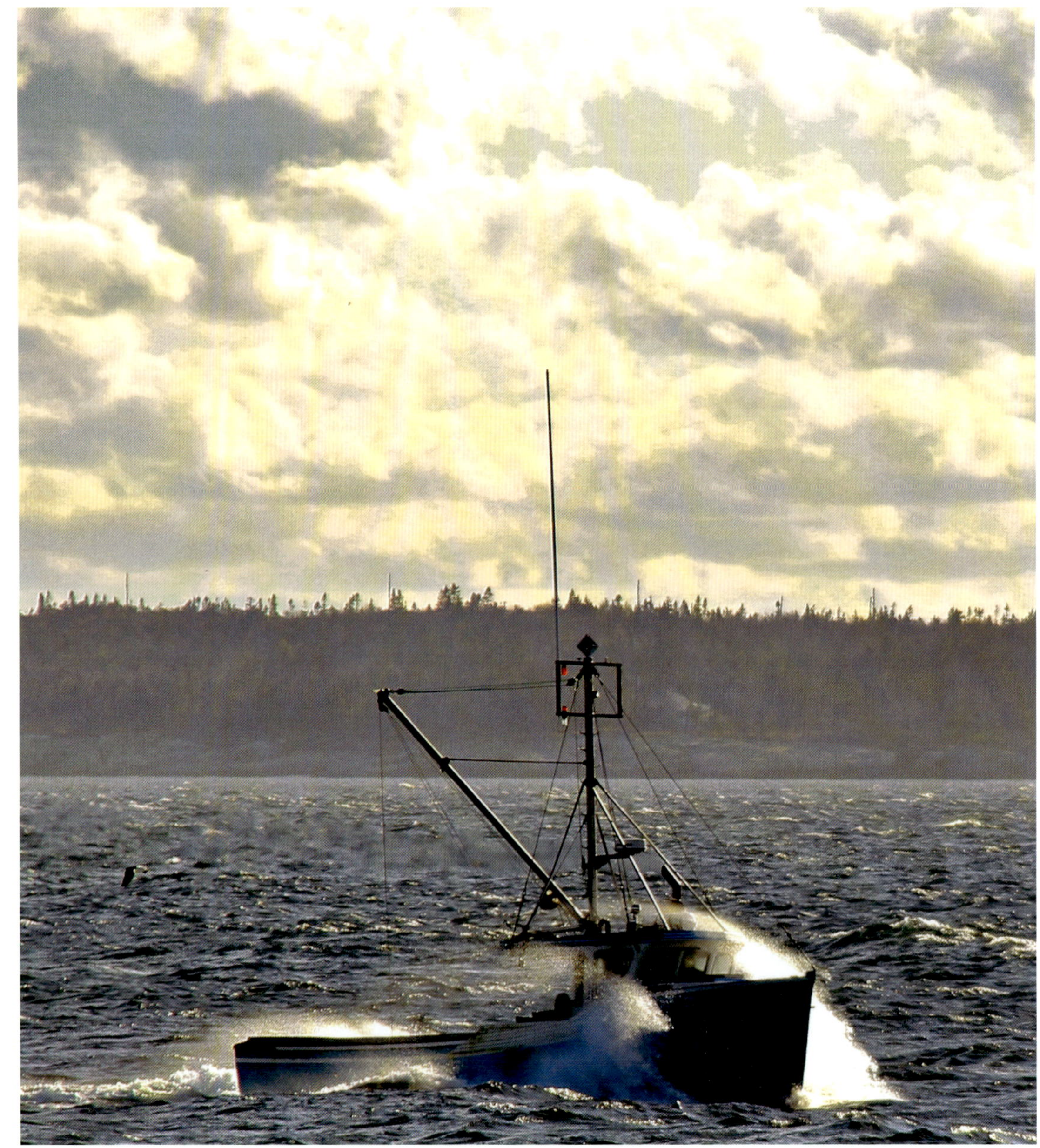

Another hard, long, risky day making a living at sea.

CLAWS

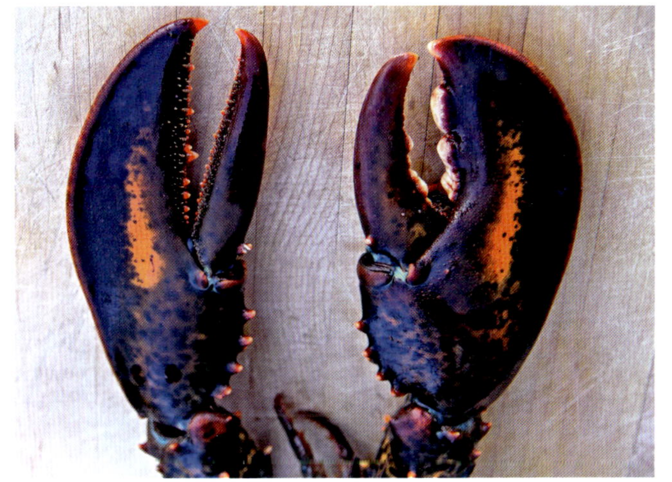

Both claws have a specific function. In this case, the lobster is right-handed: the crushing claw is on the right and the shredder claw on the left. If reversed, the lobster is left-handed.

These days, lobster traps are hauled up with the help of a hydraulic winch. One by one, each trap lid is opened and cleared of seaweed and bycatch. The old bait is removed and replaced. Spent bait is never thrown overboard—a satiated lobster is the last thing a trapper wants in his fishing grounds. Rather, the bait is placed in buckets to be emptied at sea for the gulls to enjoy once the boat is back in port.

When a lobster is landed, a fisher must carefully take it out of the trap to avoid being "bitten." Handling a livid lobster is cause for caution: the larger the lobster, the graver the potential injury. A lobster's claws can reach a radius of 320 degrees—40 degrees doesn't leave much room for a heavily gloved hand to pull a lobster out of a small trap opening. Once a lobster grabs hold, the first impulse is to try to shake it off. This makes the lobster even angrier. But forget about amputating its claw: when disconnected from the lobster's brain, its pincher can retain a vice-like grip for two to three minutes! The best thing to do is to *try* and relax your hand. This will send a signal to the lobster that the threat has passed, and it will hopefully let go within ten to twenty seconds.

Before fishers give any thought to banding the lobster's claws, its carapace must be measured. Using a specialized ruler, fishers check to see whether the lobster meets minimum legal size requirements. If too small, the lobster is thrown back to sea. If it meets size regulations, the lobster's claws are fitted with a strong, thick rubber band and it is placed in a large

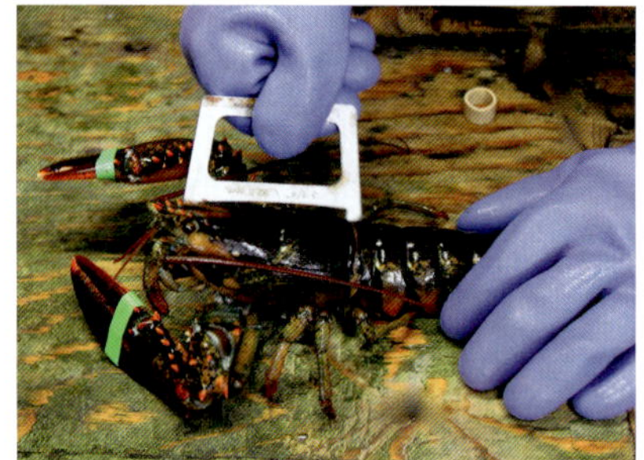
Measuring from the lobster's eye socket to the tip of the thorax to ensure it's legal (market) size.

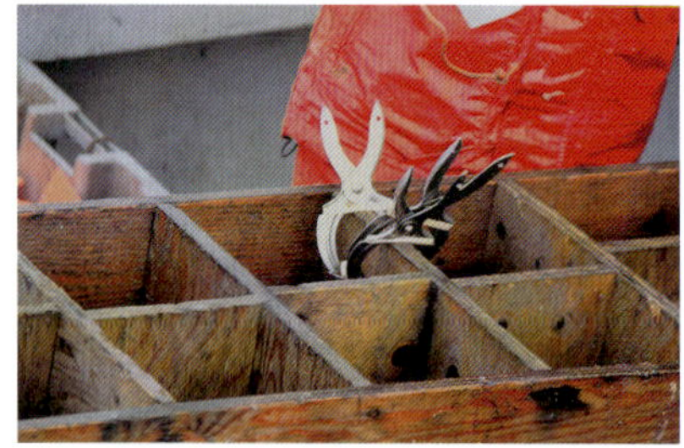
Some boats carry a "banding table" to keep lobster separated until their claws are banded.

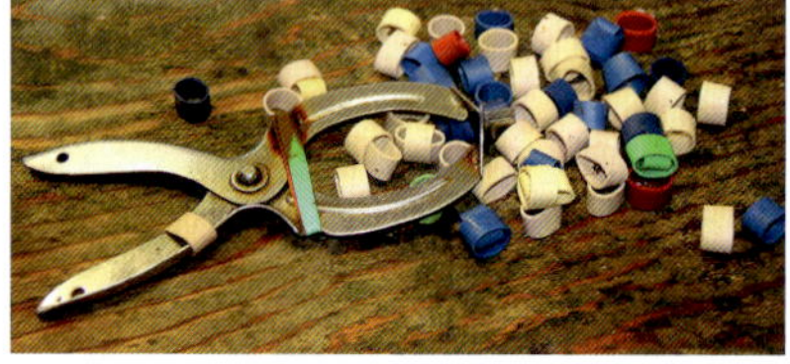
A claw-banding tool and elastic bands.

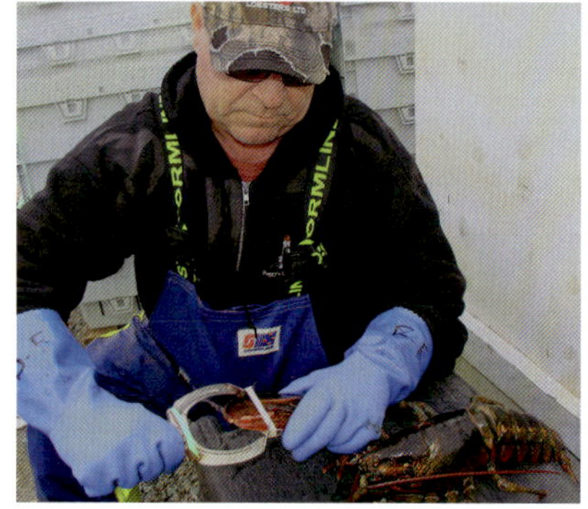
Banding a lobster's claws. Large lobster get double- and even triple-banded.

crate destined for the pound. Very large lobsters—those weighing 3 pounds or more—require at least three bands per claw.

The reasons for keeping lobster claws secured shut are many: First, for the safety of the harvesters. Second, lobsters crowded in a crate will injure each other, and the value of a lobster is gravely diminished if it is flawed in any way. Third, lobsters are very territorial, and un-banded lobsters will cling to each other like Velcro, making it impossible for harvesters to handle or separate them safely.

ROPES

Harvesters must also, at all times, be mindful of loose rope on the deck—and there is a lot of it. Entanglement or tripping could easily throw a person overboard, and just a minute or two in the frigid Atlantic waters during the winter months can be fatal. Trap ropes vary in length depending on the depth of the sea bottom. Fishers can tell the length of the rope needed by the size of the rope coils they have prepared (small rope coils for shallow waters, large coils for deeper waters). Depth sounders are immensely useful in determining appropriate rope length while also keeping the tides in mind.

Using the right length is important because if the rope is too short, the trap marker (buoy) will vanish under

Canners Versus Market-sized Lobsters

The minimum lobster-size requirement is an important conservation practice. The length changes periodically, depending on the health of lobster populations. These days, the average range for carapace length is 3.25 inches. Two types of lobsters are landed: "canners" and "market-sized" lobsters. Canners meet size regulations but are of a lower grade: it might be soft-shelled, have at least one claw missing, or have some other flaw. Most consumers of whole live lobsters prefer those between 1.5 to 3 pounds. Perfect little lobsters that weigh less than that often end up as "canners" because there is less demand for them.

water. If the rope is too long, the buoy may drift to shore at low tide and get snarled between rocks. Furthermore, an elongated floating rope can become a navigation hazard. When a vessel's propeller accidently gets wrapped in rope, the boat becomes incapacitated. This is more than a little inconvenience. Another nearby vessel has to be called in to tow it ashore. Once docked, a diver is required in to cut the rope away under water. Another solution is to pull the boat up a skid at high tide and wait for the tide to drop so that the rope can be removed manually before the tide rises again. To avoid all the potential drama of rope–propeller entanglement, most large fishing vessels are fitted with metal propeller cages.

ROCKS

And then there are the rocks. It is said that a good fisher knows the sea floor like the palm of his hand. But everything has more buoyancy in cold seawater: even the largest boulders can shift during a storm, and the low tide that follows can be most unforgiving to an inshore fishing vessel. Resting just below the ocean's surface are invisible rocks that can become a serious threat. A punctured haul and even a small crack can result in a "mayday" call.

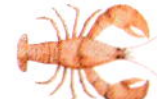

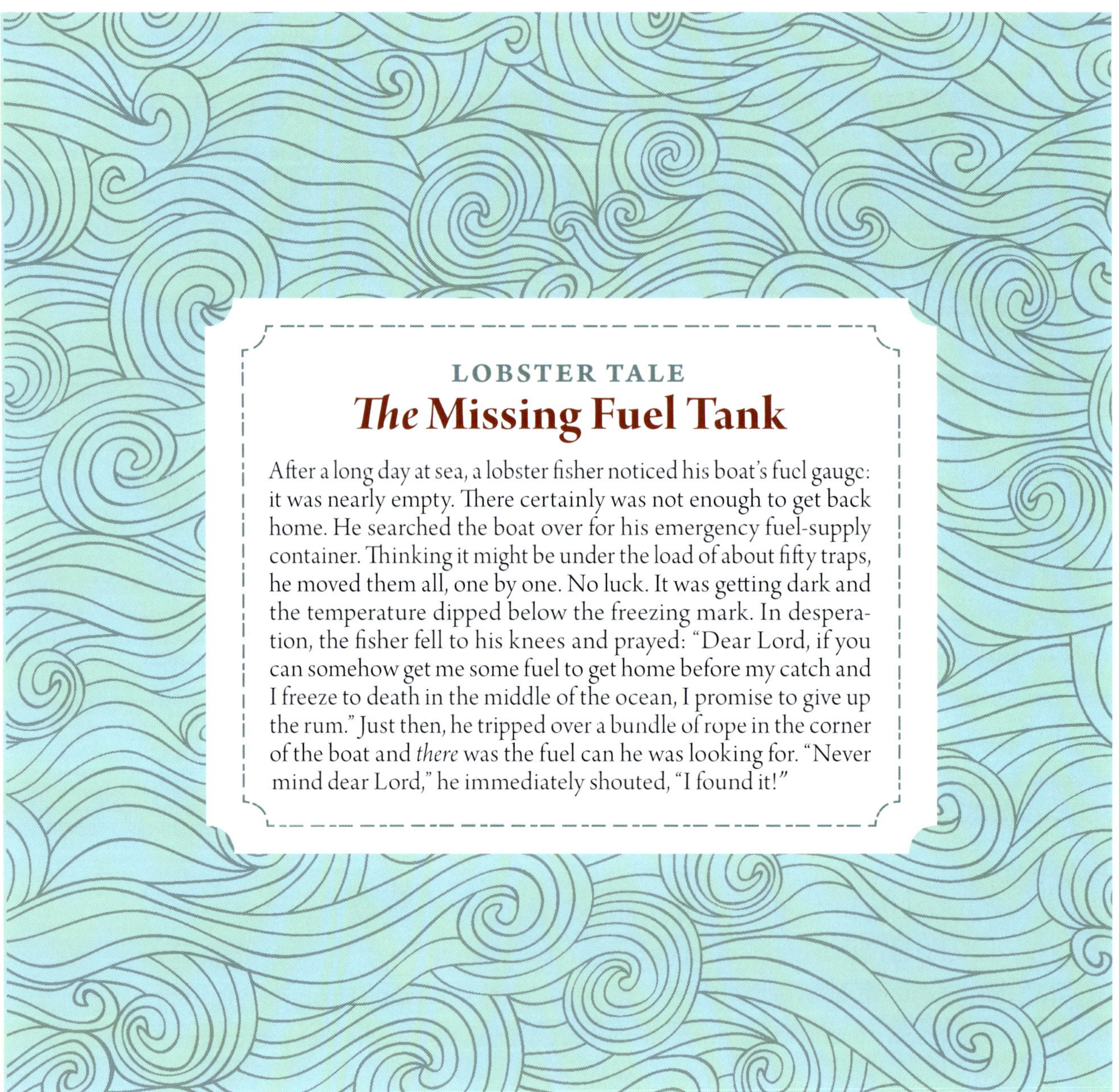

LOBSTER TALE

The Missing Fuel Tank

After a long day at sea, a lobster fisher noticed his boat's fuel gauge: it was nearly empty. There certainly was not enough to get back home. He searched the boat over for his emergency fuel-supply container. Thinking it might be under the load of about fifty traps, he moved them all, one by one. No luck. It was getting dark and the temperature dipped below the freezing mark. In desperation, the fisher fell to his knees and prayed: "Dear Lord, if you can somehow get me some fuel to get home before my catch and I freeze to death in the middle of the ocean, I promise to give up the rum." Just then, he tripped over a bundle of rope in the corner of the boat and *there* was the fuel can he was looking for. "Never mind dear Lord," he immediately shouted, "I found it!"

"It happened in the summer of 2011. The fishery had only been opened for a week. My son, my brother and I worked out of a wooden boat at that time. Hard to tell what happened, but a plank let go. It could have been a rock that caused it. We'll never know for sure.

We were a good ten miles out. In a mad panic, all three pumps were set at once, then we bolted for the wharf as fast as the boat could go. The pumps couldn't keep up at all. Thank goodness for phones. I knew my wife, Linda, was fishing lobster about eight miles away from us. I made a mayday call to her. I knew she could get to us faster than any rescue vessel.

When she finally reached us, our feet were under water. We got on her boat just in time. By then my boat was filled with water; only the cabin was showing. Wooden boats don't completely sink because wood is buoyant, but we would have been over our heads in no time if it wasn't for her.

We tied a rope to the partly submerged boat and managed to pull her to shore. I didn't really fear for my life because I knew Linda wasn't far, but my boat was a total loss."
– Donald Leblanc, lobster fisher, Bas-Cap-Pelé, New Brunswick

RIVALRY

Lobster fishers are as territorial as the crustaceans they chase when it comes to who fishes where. Each has a *claimed* traditional fishing ground. The boundaries are not clearly drawn; they are a matter of perception. With the race to get to the most prolific spots, tempers are short and the possibility of collision high. A boat loaded down with heavy traps past the gunnels is not easily brought to a halt nor can it swiftly turn without the risk of listing. Rest assured, scores will be settled in time when feathers get ruffled on Dumping Day. There are unwritten laws at sea.

Lobster harvesters unashamedly admit to dealing with issues of perceived encroachment over *their* fishing grounds in their own ways. Suspect mavericks face sabotage to their gear. If

Mayday!

The term "mayday," incidentally, has nothing to do with the month of May. It is a word that dates back to the seventeenth-century wars at sea between the British and the French. *"Venez m'aider!"* ("Come to my aid!") was an all too common cry heard from sinking French ships. From a distance the call that stood out was *"m'aider!"* The expression became anglicized over time and is now a standard naval and aviation distress call, indicating the need for urgent rescue. The mayday call is always made three times in a row to ensure it is well heard over the sound of chaos at sea or in the air.

you ever come across a buoy attached to a clean-cut rope end, it can only be the result of two things: Either it was cut because it got stuck in rocks or another boat owner didn't like where your trap was dropped. And then there is the ongoing sneaky suspicion that someone has been secretively emptying another's trap of valuable lobsters on a dense foggy day or under the cover of night. This, especially, has been the source of many family and community feuds. Outright fights do break out occasionally and hard feelings can live on for generations. Once in a while someone *snaps,* as was the case in Petit-de-Grat Harbour in Cape Breton, Nova Scotia. On June 1, 2013, there was a high-profile case of a fatal shooting at sea that became known in the media as "Murder for Lobster."

Most harvesters know not to take the law in their own hands when theft is suspected. Where stealing is severe, the DFO can conduct a sting operation to catch the offender red-handed. With today's technology, it is possible to microchip a lobster and place it in the parlour of a trap suspected of being emptied on a regular basis by a poacher. The chipped lobster has no possibility of escape unless removed manually. A DFO officer checks the trap in the wee hours of the next morning. If the lobster is missing, it is determined to be the result of theft.

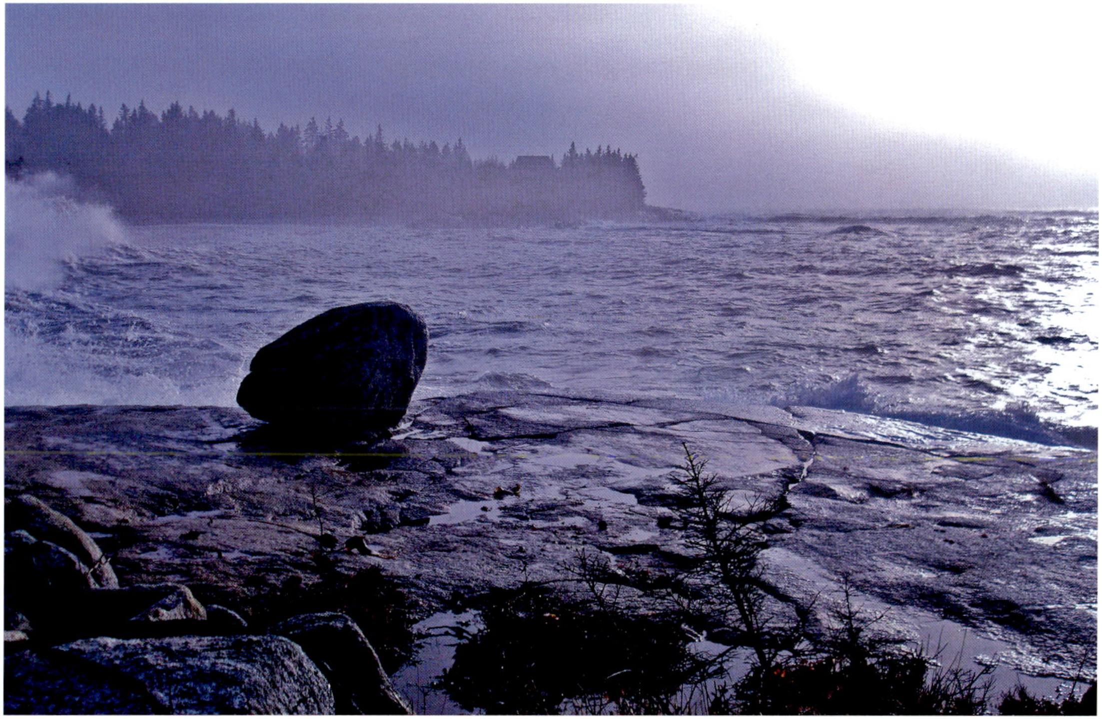

Although granite is among the heaviest of all rock types, it can still roll around in turbulent waters. A storm lifted this two-ton rock from the sea bottom onto this high ledge in St. Margarets Bay, Nova Scotia, during Hurricane Juan in September 2003.

One particular lobster harvester in St. Margarets Bay, Nova Scotia, who was the victim of chronic poaching for many years, estimates to have lost as much as $65,000 in lobster. He was greatly relieved to have it come to an end when DFO located a chipped lobster on the boat of the suspected thief. The culprit was charged and his fishing license was suspended. He also had to pay a $4,000 fine. From that point on, lobster catches in St. Margarets Bay improved substantially! Today, DFO officers have the right to board any boat any time with a lobster-chip scanner.

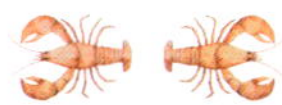

Despite all of the pressures and risks involved, thousands of Atlantic Canadians choose to earn their living at sea. What, then, is the attraction? There are many. First, lobster has become a valuable commodity and is in demand worldwide. In 2015 the Lobster Council of Canada estimated the lobster industry was worth $1.7 million to the country's economy. Lobster is Canada's most valuable seafood export, and supply remains high. Lobster can be kept live in pounds or frozen to meet peak consumer interest. Second, the lobster fishery is one of the last remaining fisheries not run by large corporations. Each harvester owns and operates his or her own boat. *Independence* is worth venturing out to sea no matter what the dangers. Some go out just long enough to get by. Other more highly motivated lobster fishers keep up the hard work as long as humanly possible to get ahead—and many do. When asked why they choose to earn a living trapping lobster, the most common answers will be along the lines of, *You get to be your own boss on the water, and being at sea is way better than being stuck in an office or a warehouse, no matter what the sea throws at you.*

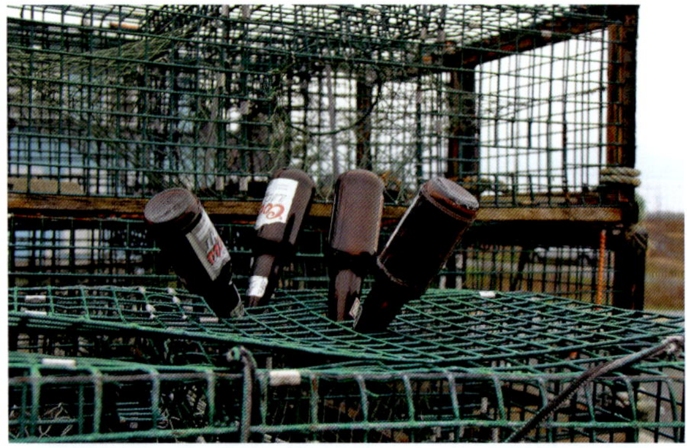

A few cold ones to celebrate a good catch at the end of a hard day's work.

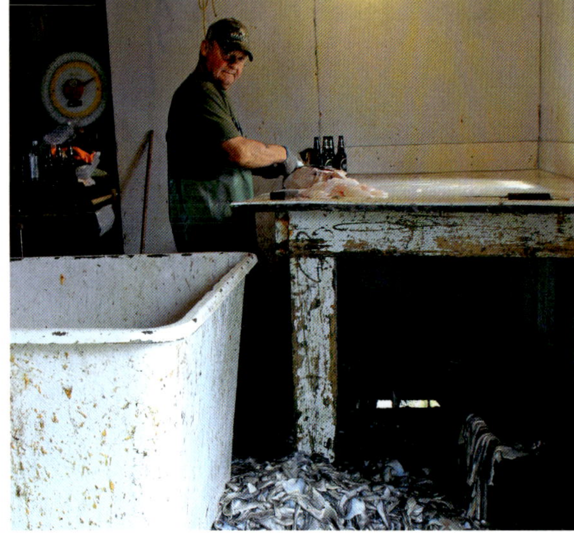

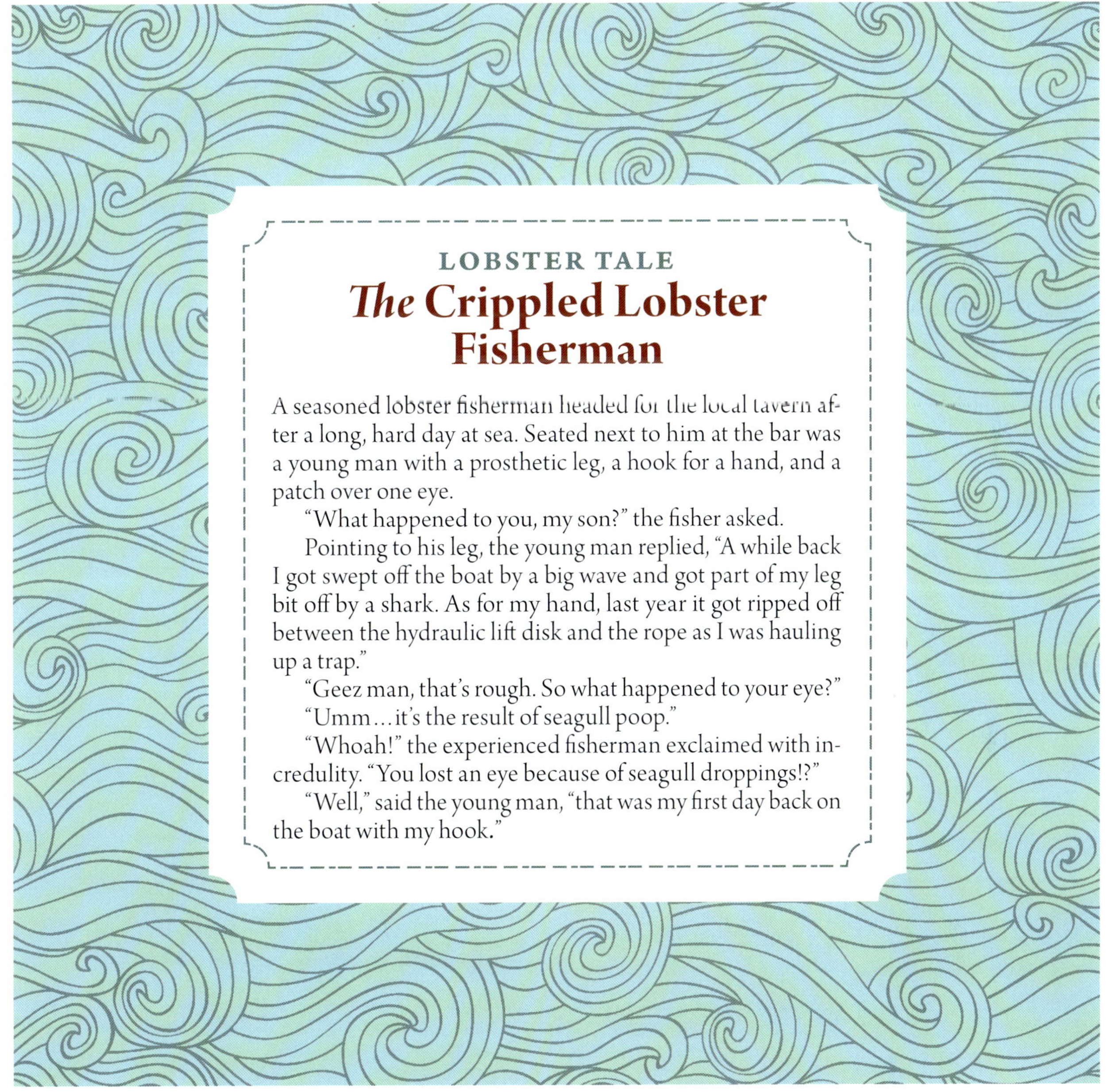

LOBSTER TALE

The Crippled Lobster Fisherman

A seasoned lobster fisherman headed for the local tavern after a long, hard day at sea. Seated next to him at the bar was a young man with a prosthetic leg, a hook for a hand, and a patch over one eye.

"What happened to you, my son?" the fisher asked.

Pointing to his leg, the young man replied, "A while back I got swept off the boat by a big wave and got part of my leg bit off by a shark. As for my hand, last year it got ripped off between the hydraulic lift disk and the rope as I was hauling up a trap."

"Geez man, that's rough. So what happened to your eye?"

"Umm…it's the result of seagull poop."

"Whoah!" the experienced fisherman exclaimed with incredulity. "You lost an eye because of seagull droppings!?"

"Well," said the young man, "that was my first day back on the boat with my hook."

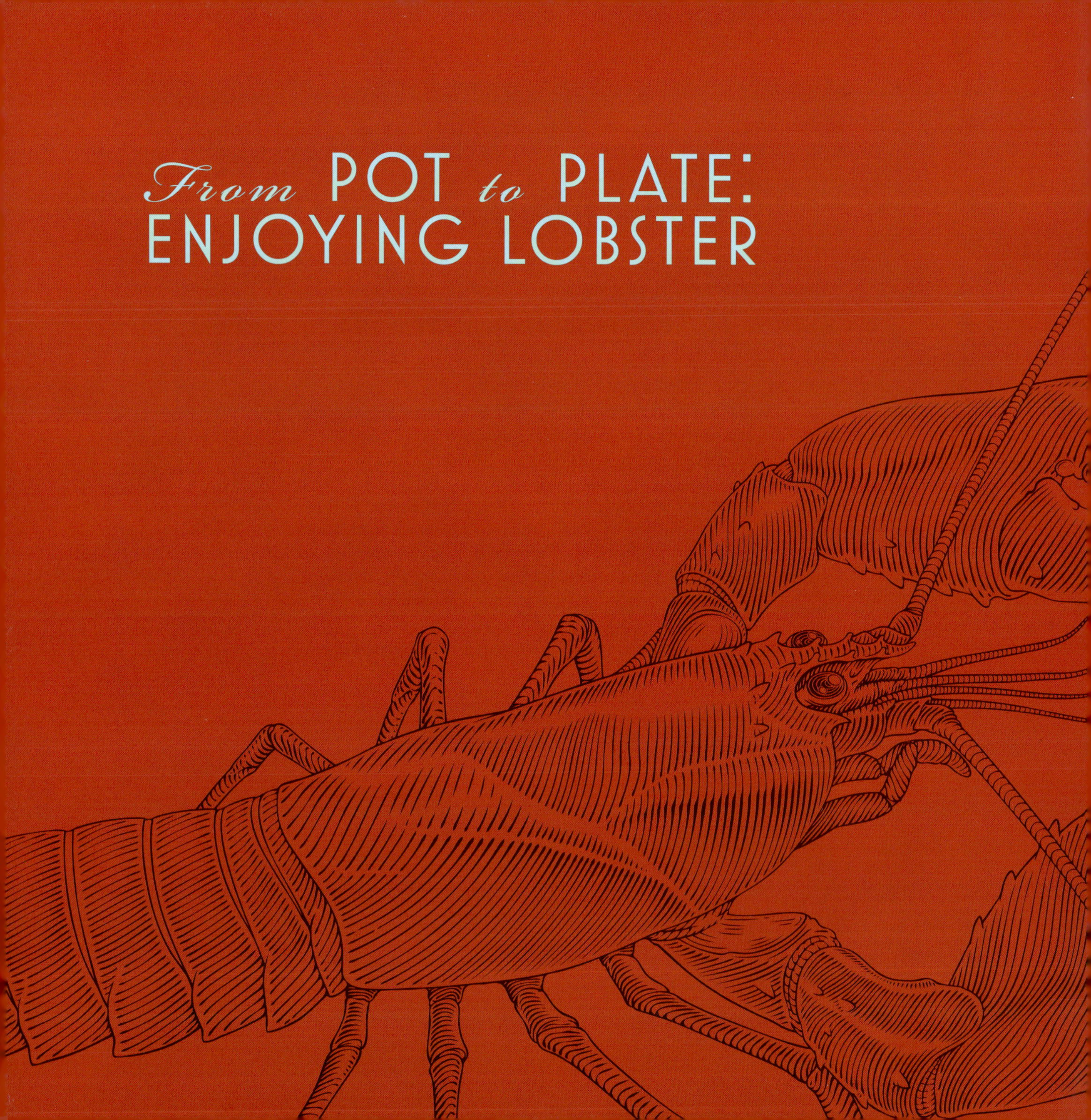
From POT to PLATE:
ENJOYING LOBSTER

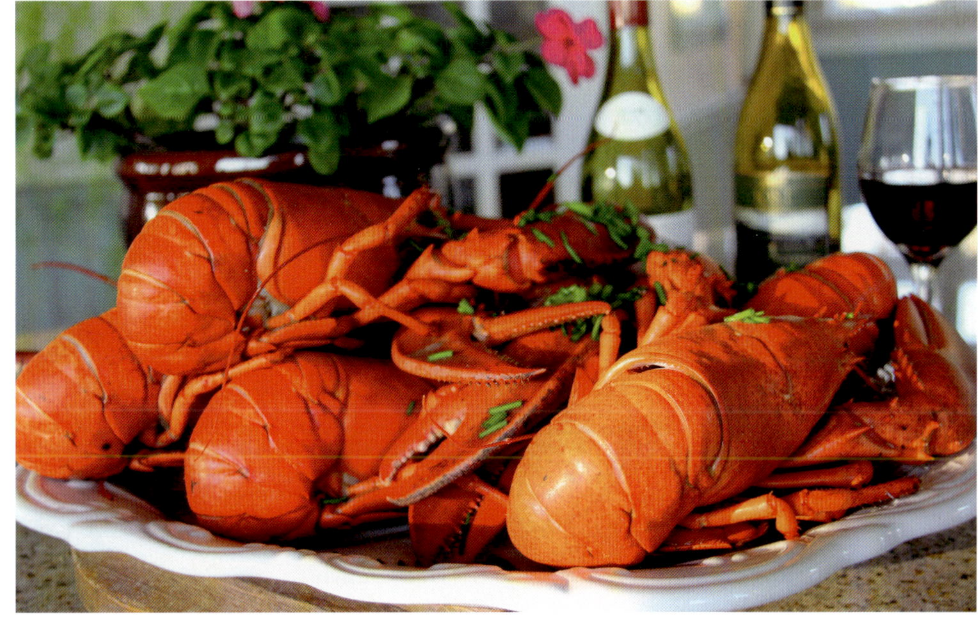

Once cooked, lobster's brilliant red colour and aroma of salt spray stimulate the appetite and compel us to get crushing.

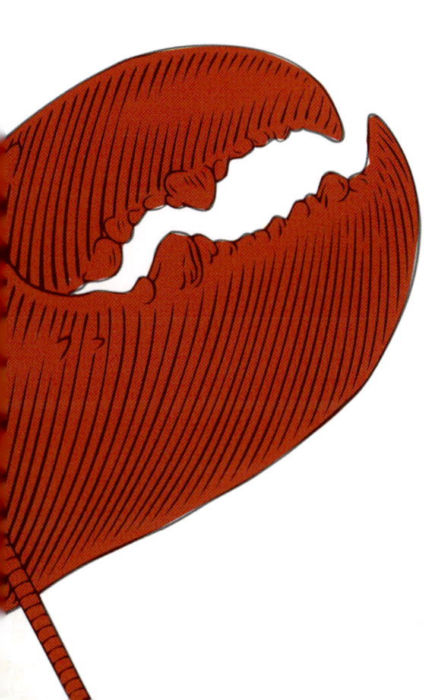

The Lobster Experience

Lobster has become the ultimate culinary experience in Atlantic Canada. People from all corners of the earth continue to stream to our coasts to feast on fresh-caught wild lobster. They arrive with a pilgrimage-like fervour, as if they're visiting the birthplace of a deity. Folks "from away" who have a meal of Atlantic lobster for the first time spread the news quickly, calling it the best seafood experience they've ever had. Many return to the same spot year after year for this very reason. Americans have access to plenty of lobster from the coast of Maine, but after visiting Atlantic Canada and enjoying a meal of locally caught lobster from the frigid waters of the North Atlantic Ocean, they will tell you again and again, it is the best.

Lobster is celebrated in this region culturally and commercially. Although in supply year-round, lobster sales spike during our holidays and festivities: Thanksgiving, Christmas, New Year's Eve, Canada and Victoria Day long weekends. We might also order lobster for a special milestone: a birthday, a wedding anniversary, a promotion,

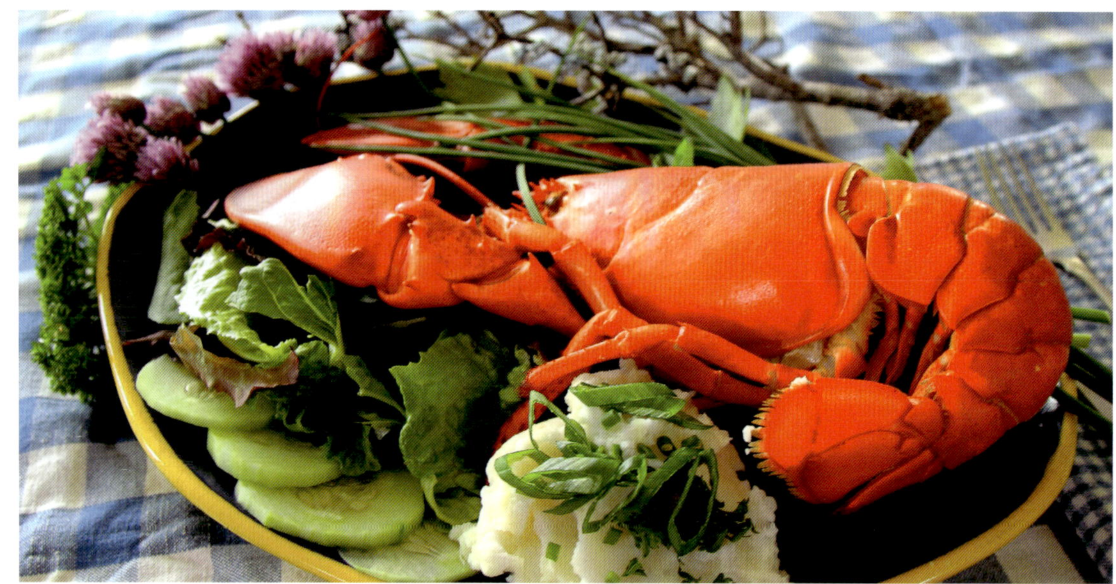

Home-cooked lobster makes an easy and relatively affordable meal.

or as a way to ring in retirement. How about a staff party, a banquet, or following a hockey game? What better way to welcome a newcomer to the neighbourhood—and to Atlantic culture—than to share a meal of lobster? Whatever the occasion, when planning for a *really* special, out-of-the-ordinary meal, what comes up first is not turkey, roast beef, or ham—it's lobster.

Natural, free-range, and low in calories, lobster is an excellent source of flavourful protein from the sea. Legumes, quinoa, and tofu, although just as natural and high in protein, don't draw a crowd in quite the same way. Lobster has status, prestige; it is a symbol of success—and at times, even excess. It is not eaten only to mark special occasions, oh no. Folks eat lobster just for the sake of enjoying a satisfying meal fresh from the sea. The

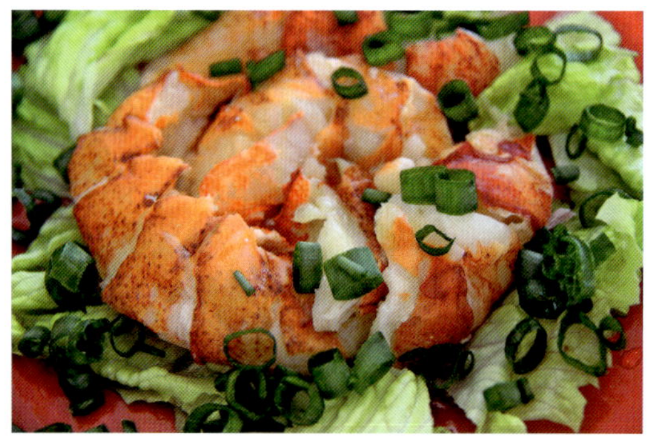

Lobster is a wholesome, natural source of protein with outstanding health benefits.

Nutritional Value of Lobster

Lobster is the ultimate free-range, wild-caught health food. It is lean, high in protein, and rich in many essential vitamins and nutrients. It has fewer calories, fats, and cholesterol than lean beef, roasted skinless chicken breast, and poached eggs. In fact, a meal of lobster is nearly fat-free; it's the butter, dipping sauces, cream (in lobster chowder), and dollops of mayonnaise (lobster rolls) that bump up the fat content. With that in mind, consumers of lobster can enjoy it straight from the shell or indulge in luscious preparations of the succulent crustacean guilt-free.

Besides being a heart-healthy protein that is low in fat, calories, and cholesterol, according to the University of Maine's Lobster Institute lobster is also high in amino acids, potassium, and magnesium, vitamins A, B12, B6, B3 (niacin), and B2 (riboflavin), calcium, phosphorous, iron, iodine, and zinc.

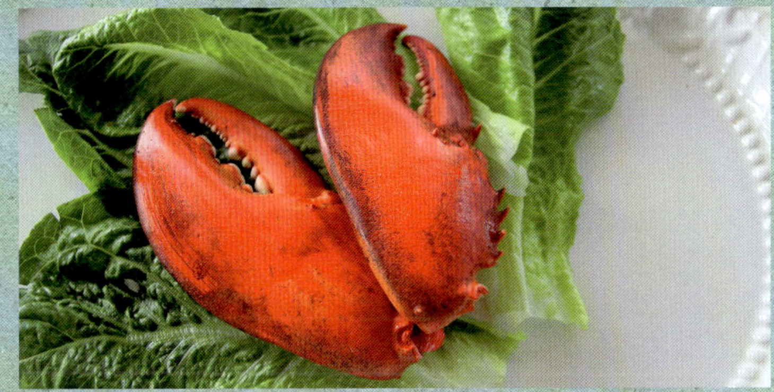

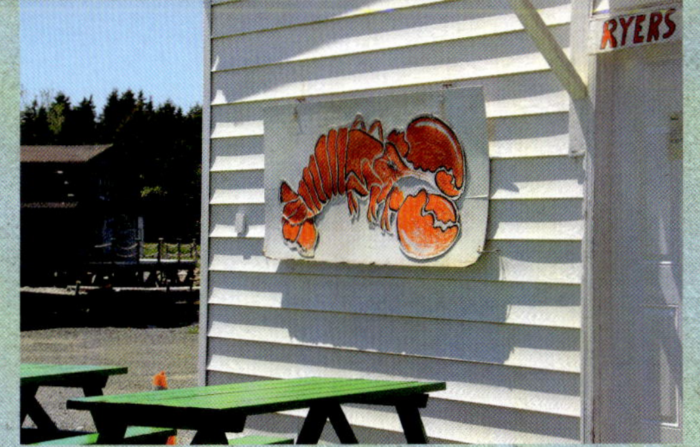

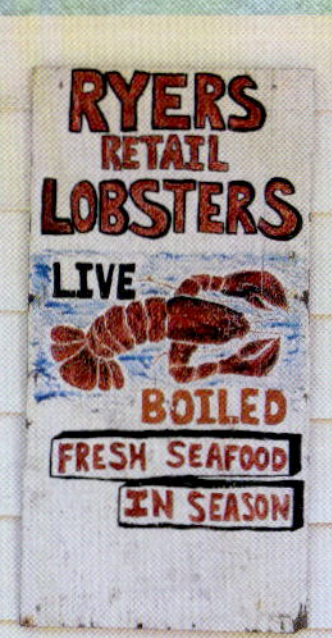

Ryers local lobsters can be enjoyed outside, overlooking the very waters where they were trapped, or as takeout, live or cooked.

Food item*	Calories	Cholesterol	Fat (mg)
Lobster (boiled)	98	72	0.6
Chicken breast (skinless, roasted)	165	85	3.6
Beef (lean)	216	86	9.9
Whole egg (poached)	149	423	10.0

* *Per 100g portion*

The Cost of Lobster

Here is approximately (market value fluctuates) what a 1 1/2-pound Atlantic-caught lobster dinner can go for:

- Gourmet, mess-free, shell-off arrangement of lobster in a chic European restaurant: $150.

Cooked whole lobster in the shell

- Toronto's CN Tower seafood bar: $75
- Halifax, Nova Scotia, waterfront seafood restaurant: $40
- Sou'Wester Restaurant, Peggys Cove, Nova Scotia: $30

No-frills-by-the-wharf at Ryers Lobster Retail, Indian Harbour, Nova Scotia

- Freshly cooked whole lobster in the shell eaten on the spot or takeout: $13.50
- Cook-it-yourself live lobster: $12.50
- Live 1 1/2-pound lobster bought directly from a fisher*: $9

*On average, a fisher receives about $7.50 per lobster from a wholesaler. In the 1920s a fisher got 10 to 15¢ for each pound of lobster.

"lobster experience" has become a global phenomenon. It takes many forms and pricing varies astronomically depending on *where* and *how* you dine.

Obviously, the closer to the source you buy your lobster, the better the price. And if you buy locally, you'll be consuming the freshest free-range lobsters imaginable. In contrast to the fine-dining lobster experience, but no less indulgent or delicious, is to eat the crustacean in the most primitive way. Lobster grants us permission to get down and dirty for the prize inside. Many would agree that the best lobster experience is when you drop all table manners and etiquette and dig in, bib on, mallet for crushing, and rudimentary tools for prying, pinching, and pulling. You can make sucking, slurping sounds, lick your fingers, and drip melted butter across the table without guilt or shame. That's the quintessential Atlantic-style lobster experience. Rah-rah for Atlantic cold-water lobster enjoyed where it is caught!

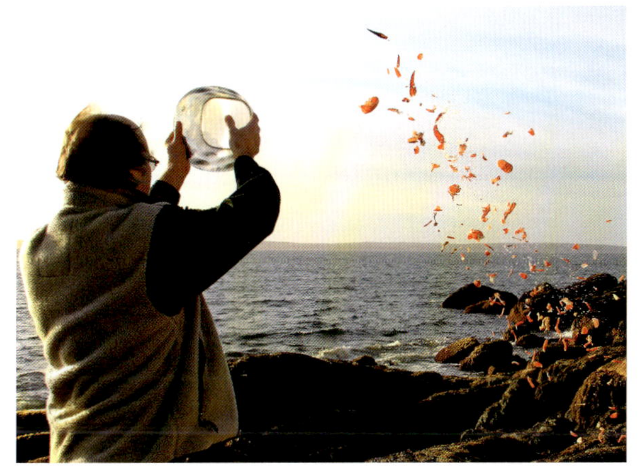

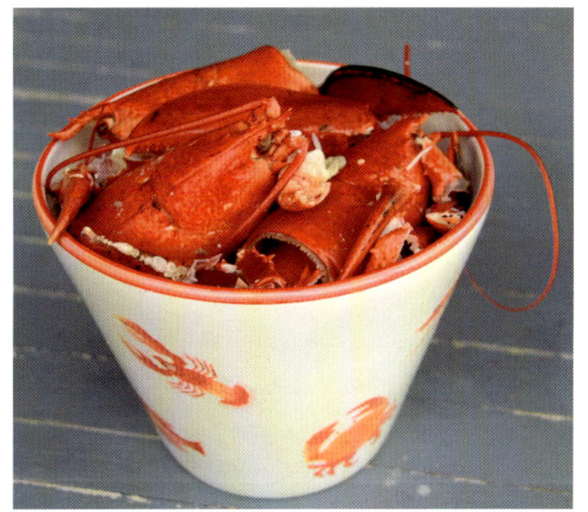

What to do with those shells? They can go in the compost—or, better yet, save them for lobster bisque (page 121).

The "How to Cook a Lobster Humanely" Debate

The act of boiling live lobster for consumption has been a point of contention for some time. There have been many protests against it, particularly in the United States, and celebrities—such Mary Tyler Moore, Paul McCartney, and Brigitte Bardot—have spoken out against cooking live lobster, drumming up attention to the issue. The good news about these kinds of events is that they bring concerns about animal welfare to the forefront. The fundamental debate here is whether lobsters are as sentient as our pets.

Many marine biologists suggest lobster don't experience pain to the same degree as vertebrates, that they have a primitive nervous system on par with insects. Lobster *are* invertebrates: they lack a spinal cord. And a lobster's brain is about the size of a pinhead, so its nervous system is nowhere near as complex as ours. There is a tendency for humans to anthropomorphize creatures that are not well understood. But when concerned consumers learn that lobsters do not care for their young, do not display affection toward their kind, and will even prey on their own, it becomes clear these are creatures of a different type.

Furthermore, lobsters have no ability to express emotions. Unlike your cat or dog, a happy lobster looks as fearsome as a mad lobster. They can also self-amputate (see page 28),

Lobster will self-amputate when water gets too close to freezing, too warm, or low in oxygen, and to escape when a predator catches it by a claw.

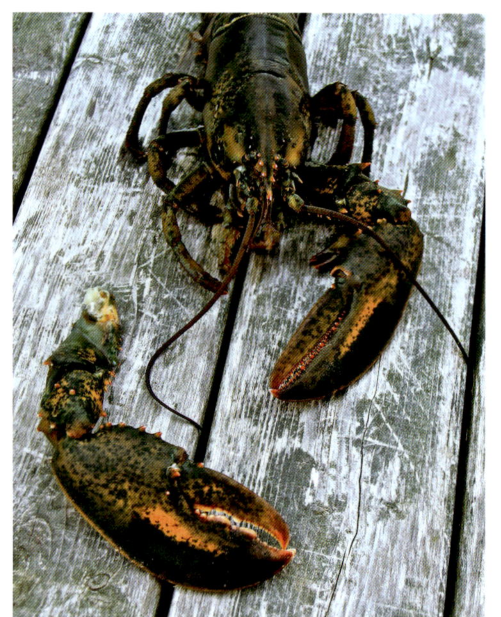

voluntarily dropping a claw without any apparent signs of discomfort. Touch an earthworm with a stick and you'll get a lot more reaction. Lobsters don't panic when trapped the way a wild land animal would. Even bees and flies try desperately to escape when trapped behind glass. A lobster, on the other hand, often just settles in a corner of a trap and relaxes. Experts believe this behavior suggests lobster perceive the trap as a source of easy food. They don't seem to know they are trapped until a gloved hand pulls them out.

Where pain is concerned, there is consensus on this matter: a quick death is always more humane than a prolonged one. A more constructive debate is whether a quick death by the pot is better or worse than a "natural" lobster death. In the wild, lobster die due to disease or chronic infection from a wound (a result of fighting other lobster—usually competing males). Others die from a moult gone wrong. These ailments can continue for extended periods of time—weeks, even months.

A lobster's most common natural cause of death is to be eaten by a predator. Juvenile lobsters are most susceptible, but market-sized lobster, too, risk being devoured by large, scavenging fish, mink, otter, or seal, especially at the soft-shell stage. The worst culprits are gulls. They are ravenous. Herring gulls feed mostly on crabs, but now and then they land a lobster trapped in a shallow tidal pool. Getting pecked to death by a gull is no easy way to go. There is a misconception among some nature enthusiasts that all things *natural* are beautiful, pure, and kind. A death by the pot is far less tortuous than a natural death in the

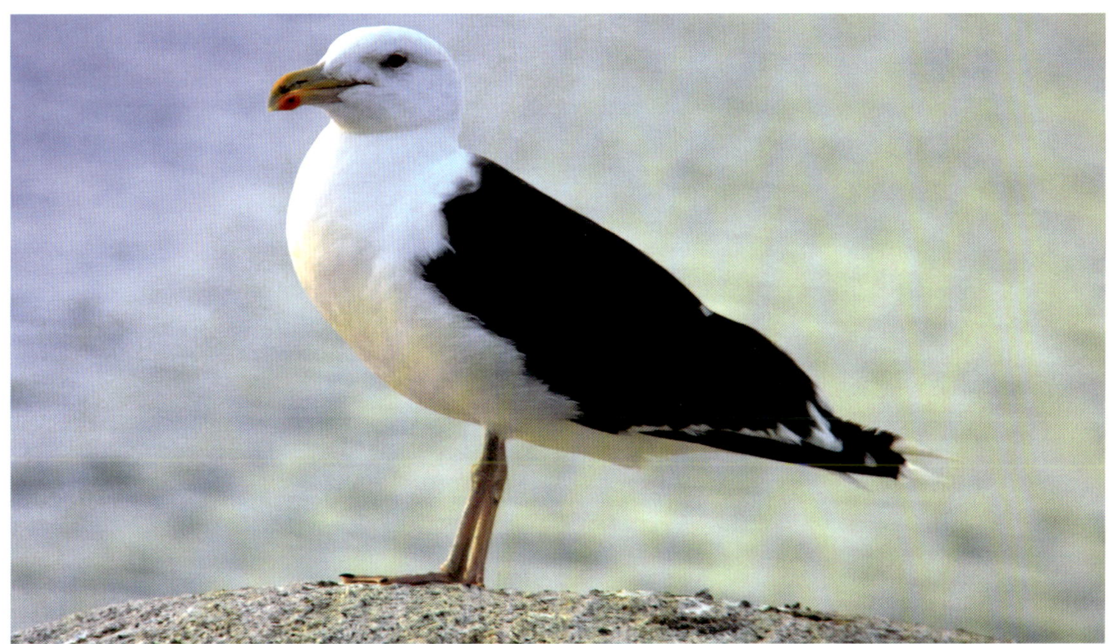

The largest of seagulls, the great black-backed gull, is a common seabird easily capable of killing and eating a 1-pound lobster in one sitting.

wild. Once the lobster hits boiling water head first, it is over in seconds.

The best way to cook a lobster humanly is to do it quickly in a good size pot of seawater or salted water with an audible, strong boil. When done correctly, there should be no more than one or two tail flicks in the first couple of seconds. While boiling a lobster to perfection takes about the same amount of time as it takes a seagull to kill it, death occurs within seconds in the pot. Putting a live lobster on the BBQ, in a steamer, or in the oven is arguably a much slower death and one that is frowned upon by concerned animal lovers.

Little Lobster Laugh

Why do lobsters make such bad businesspeople?

Because they always end up *in the red.*

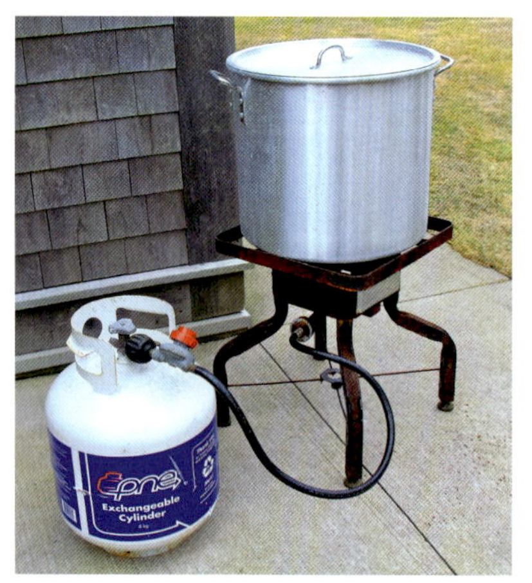

Cooking more than two lobsters at a time is best done in a large, propane-heated pot of water or a turkey roaster over two stove burners.

Instructions for Cooking Live Lobster

To keep your lobsters live and healthy until cooking time, store them in a cool place like the refrigerator, or wrapped in newspaper on ice. It's important to keep live lobster moist. Achieve this by keeping them covered with damp paper in a closed box. If stored properly, lobster can live for up to two days out of water. But for best results, cook live lobster as soon as possible.

It is important to ensure there is sufficient space and water in the pot to submerge the lobster completely and that the water, preferably seawater, has come to a full rolling boil before the lobster go in. If using tap water, add 1/3 cup coarse sea salt for each gallon of water. Remove rubber bands from claws using scissors and drop lobster in head-first, then cover the pot immediately with a lid until a full boil resumes, about 3–5 minutes.

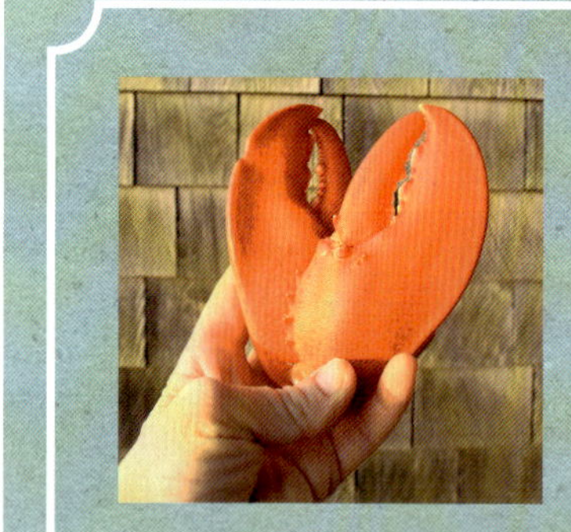

Red Lobster Demystified

Scientists from the school of chemistry at Manchester University in the UK discovered what makes lobster turn red when cooked. They have found the change in colour is due to a very stable chemical in the lobster's shell called astaxanthin. On its own, it is a red compound resistant to the boiling point of water. A live lobster's shell has many other colour pigments, such as yellow, green, and blue, which are overridden by a dark protein called crustacyanin (the lobster's camouflage). This is why a live lobster is so dark. Heat causes the other pigment compounds to break down, leaving behind the brilliant red properties of astaxanthin.

Maintain a gentle boil. Time of boil depends on the weight of the lobster.

- 1-pound lobster: 15 minutes
- 1 1/2-pound lobster: 20 minutes
- 2–3-pound lobster: 25 minutes

Leftover lobster in the shell should be refrigerated on their back in an airtight container so that flavourful juices are retained. Storing them in plastic bags will result in leakage from piercing. Their shells are very sharp. If your leftover lobster won't be consumed within the next four days, consider freezing the meat. Remove the shells and place the meat and juices in good-quality zip-lock freezer bags or plastic tubs. Remove as much air as possible. Frozen lobster is best consumed within the first three months. Bon appétit!

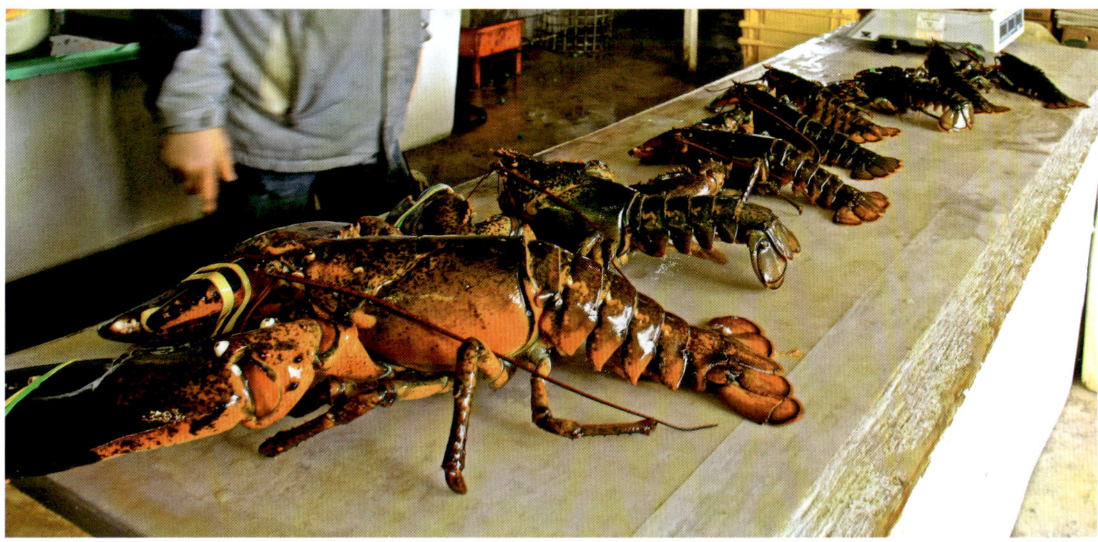

The larger the lobster, the longer the boiling time.

OLD-FASHIONED LOBSTER RECIPES *with a* CONTEMPORARY TWIST

Lobster Dishes

Much like a traditional festive meal of turkey, there will likely be leftovers following a feast of lobster. But the possibilities for serving this delicious crustacean the second time around are far more exciting and versatile. The following are fresh twists on good old-fashioned lobster favourites.

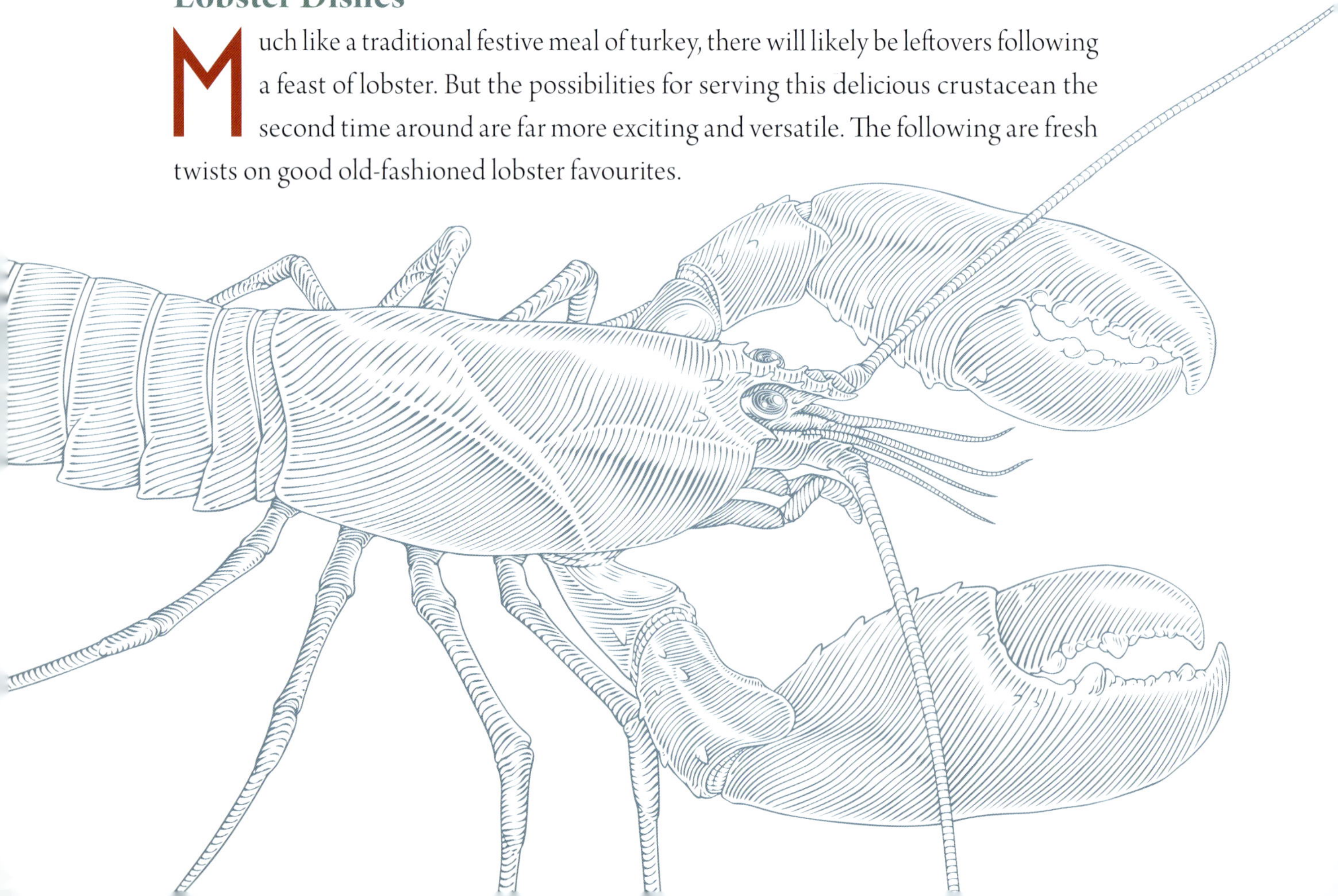

SILKY SHE-LOBSTER BISQUE

serves 4

Save those lobster shells! Any parts that have not been sucked on—the body and tomalley (pale green pudding-like organ), the shells of claws, tail, and little side legs—will serve to make a beautiful lobster stock, the base of lobster bisque. For added flavour and colour, set some bright red she-lobster roe aside. (This recipe works just as well without the roe, but you'll miss the specks of flavourful goodness.)

This wonderful French-style bisque is made in three phases: The lobster stock, a roux for thickening, and finally the velvety bisque. Cooking will take a good part of the day, so you may want to prepare it in advance and serve it the following day.

The Stock

- shells of 3–4 lobsters
- 1 clove garlic, grated
- 2 sticks celery, chopped (leaves included)
- 1 large carrot, chopped
- 1 red bell pepper, chopped
- 1/2 medium-sized onion, chopped
- 1 cube chicken bouillon
- 12 cups water

Place all ingredients in a large pot or family-size slow cooker. Bring to a boil, and then lower to a slow simmer for about 1 hour on stovetop. If using a slow cooker, you'll know the stock is at its peak when your kitchen becomes filled with an amazing aroma of the sea, after 4 or 5 hours.

Turn the heat off. Keep the lid on for an additional 45 minutes for flavours to continue releasing. Using a sieve to strain the solids out, pour stock into a large soup pot.

Continued on page 123 ▸

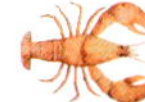

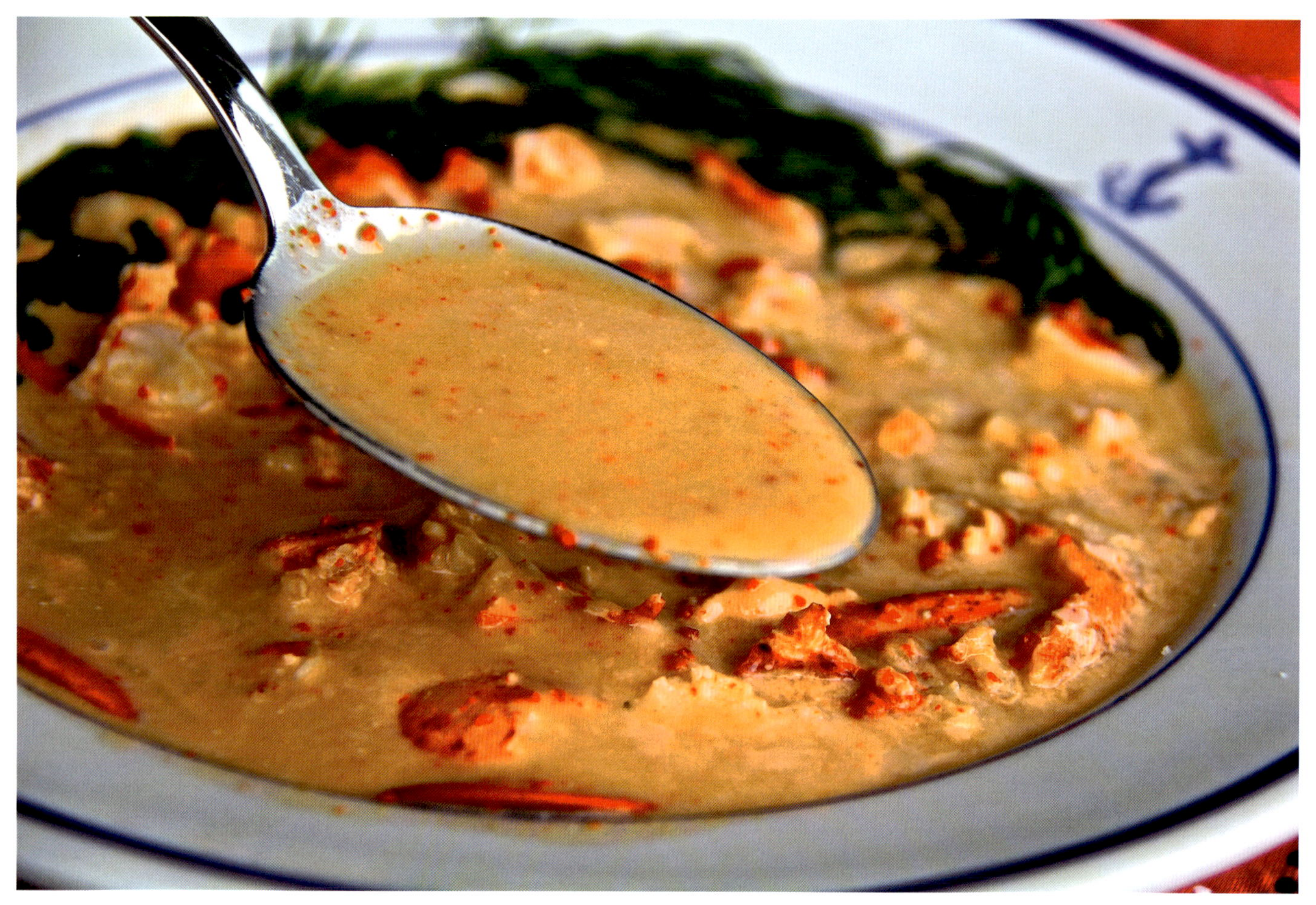

The Roux

- 4 tablespoons melted butter
- 1/4 cup all-purpose flour
- 2 tablespoons cornstarch

In a good-size bowl, combine all ingredients using a fork. Bit by bit, add heated lobster stock, then whisk together until completely blended. Bring to a boil, whisking occasionally to ensure roux doesn't settle to the bottom of the pot. Simmer 20–30 minutes until stock has thickened to the consistency of light gravy.

The Bisque

- 2–3 cups chopped leftover lobster meat
- thickened lobster stock
- 1 (170 ml) can Carnation Thick Cream
- reserved lobster roe
- paprika
- sea salt to taste
- chopped fresh chives (optional)

If using lobster roe, rinse well in a sieve and then crumble finely with a fork or with your fingers and add to the stock. For finer roe bits, use an immersion blender in the stock. Add cream and chopped leftover lobster meat. Heat the bisque but do not boil.

When ready to serve, heat bowls with boiling water. Discard the water and pour in the gorgeous, fragrant bisque with a sprinkle of paprika, sea salt, and chopped fresh chives.

Serve with fresh crusty bread.

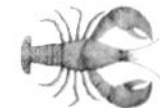

QUINOA LOBSTER ROLLS

makes 8

Lobster rolls are a classic. Adding quinoa, an ancient grain rich in protein, to the mix is a wonderful and nutritious way to stretch out your leftover lobster meat. Quinoa seems to take on any flavour it is paired with and adds a fluffy texture. Guests will wonder what your secret is for such superb, satisfying, and healthy lobster rolls.

- 3 cups cooked lobster meat
- 1 cup prepared quinoa
- 1/2 cup mayonnaise
- 2 sticks celery, shredded
- dash mustard powder
- 8 hot dog buns or croissants
- 1 tablespoon prepared garlic butter
- 4 romaine lettuce leaves, halved, centre rib removed

Chop lobster meat to smaller than bite-size pieces (scissors work well). Be sure to remove the clear cartilage inside the claws. In a large bowl, mix meat together with mayonnaise, mustard powder, celery, and quinoa.

If using hot dog buns, melt garlic butter in a non-stick skillet and then sizzle buns on each side just until golden. Fill buns first with 1/2 leaf of romaine lettuce and then add the lobster mix.

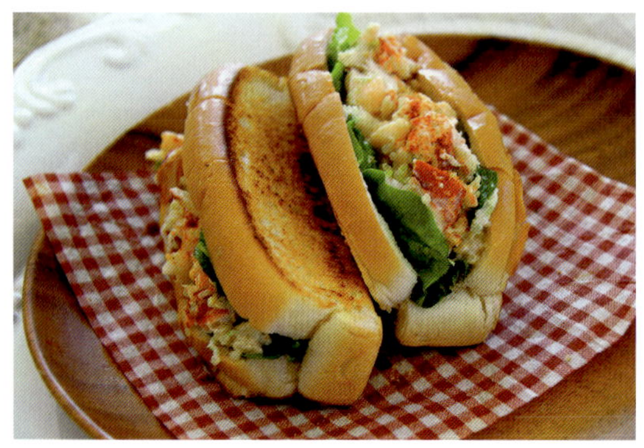

If using croissants, slit in half lengthwise (two half-moons). Spread mayonnaise on first half, topped with romaine leaf and lobster mix, and sandwich with second half.

*This lobster filling can be refrigerated for up to two days but is best enjoyed at room temperature. Take it out of the refrigerator about 1 hour before serving time. Make the lobster rolls just when every one is ready to gather at the table to avoid buns or croissants getting soggy.

LOBSTER BRUSCHETTA

serves 4 as an appetizer or snack

As beautiful as it is delicious, this is the ultimate finger food. Make the lobster mix and garlic bread slices a day in advance and all you need is 5–10 minutes to put it out fresh for an impressive, classy hors d'oeuvre.

- 2 cups finely cut lobster meat
- 1 stick celery
- 1/4 cup mayonnaise
- French baguette
- garlic butter
- 1/4 cup chive and onion cream cheese spread
- fresh baby spinach leaves, stems removed
- paprika

Slice lobster meat using scissors, making sure to remove the cartilage inside the claws. (Lobster roe may be used but not the tomalley.) Shred the celery stick, removing any strings. Add to lobster meat along with mayonnaise. Mix well.

Slice French baguette into twelve 1 1/2 centimetre–thick pieces. Butter both sides of each piece with garlic butter and place into a heated non-stick pan until both sides are golden. Let cool to room temperature.

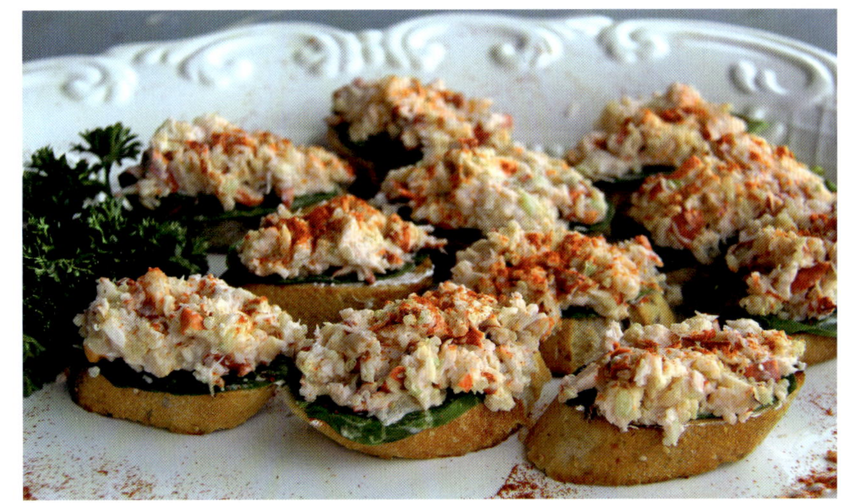

Spread cream cheese thinly on one side. Press a spinach leaf over the cheese of each slice of bread. This allows the spinach leaf to stick and prevent the lobster mixture from dampening the bread.

Top with a heaping tablespoon of the lobster mixture, and garnish with a dash of paprika.

LOBSTER RICOTTA QUICHE

serves 5–6

This is a satisfying, foolproof quiche recipe with an unforgettable hint of the sea. Lobster Ricotta Quiche rises beautifully and while lobster meat tends to be heavy, the ricotta cheese gives this quiche lightness and smoothness. Each slice is a clean pie wedge that holds up handsomely on the plate. Makes the perfect breakfast or brunch.

- 1 1/2 cups chopped lobster meat
- 1 (10-inch) piecrust (store-bought frozen for convenience)
- 1 cup chopped fresh baby spinach
- 5 eggs
- 1 (475 gram) tub ricotta cheese
- 1/2 yellow or white onion, minced
- 1/2 teaspoon sea salt
- 3/4 cups shredded cheddar cheese
- dash paprika

Defrost piecrust and place it into a 10-inch glass pie plate. Precook at 350°F (170°C) until firm, about 10 minutes. Allow crust to cool. Cover the bottom of crust with spinach.

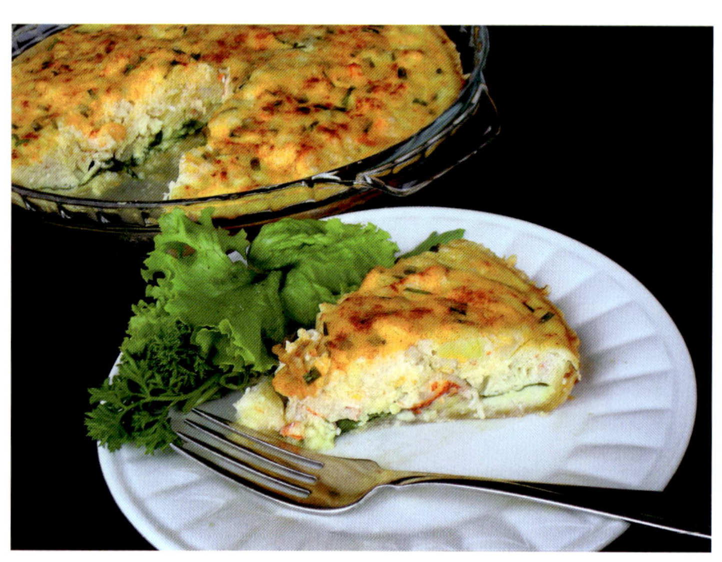

In a large mixing bowl, beat the eggs and add lobster meat, ricotta, onion, and salt. Stir ingredients together. Pour mixture over the spinach. Cover with cheddar and a sprinkle of paprika.

Bake at 400°F (200°C) for 45 minutes or until firm. Allow quiche to rest in the oven another 10 minutes. Serve with your favourite light salad.

LOBSTER AND AVOCADO CREPES

makes 6–8

This recipe requires a bit of work but is well worth the effort and can be prepared ahead of time. There are three parts: the lobster filling, the avocado filling, and finally, the crepes. Once all three items are prepared, putting the crepes together takes under 10 minutes. The creaminess of ripe avocado paired with the briny taste of lobster is a winner.

Lobster Filling

- 3 cups lobster meat
- 1/2 cup mayonnaise
- 1/2 cucumber, seeded and diced very small (2–3mm)

Slice lobster meat finely using scissors. Place meat in a bowl with mayonnaise. Add cucumber. Mix well. Let sit, covered, at room temperature while preparing avocado filling.

Avocado Filling

- 1 small clove garlic
- pinch sea salt
- 1 ripe avocado
- 1/4 cup mayonnaise
- handful baby spinach leaves
- tomalley (optional)

In a separate bowl, grate garlic and mash sea salt into it with a fork. Slice avocado in half, lengthwise. Remove pit and spoon flesh into bowl of garlic. Add mayonnaise and mash together. Remove stems from spinach leaves and chop finely. Add them to the bowl and combine ingredients.*

Let sit, covered, at room temperature while preparing the crepes.

*Incorporating tomalley at this stage works very well as the colour is compatible and adds extra lobster flavour.

Continued on page 128 ▸

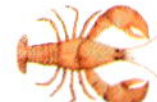

Crepes

- 2 eggs
- 2 cups white flour
- 1 1/2 cups milk
- 1 teaspoon vegetable or sunflower oil
- pinch sea salt
- paprika

Crack eggs into a large mixing bowl. Whip for a few seconds until blended. Add flour, milk, oil, and salt. Mix well until smooth. Crepe mixture should have the consistency of eggnog (add more milk if too thick).

Unless using a good-quality non-stick frying pan or skillet, use a small amount of non-stick oil. Heat pan or skillet to medium-high. Using a ladle, pour in crepe mixture to form desired circular crepe, about the size of a medium tortilla (20 cm) across. Note that regular-sized frying pans will only have room for one crepe at a time. When tiny bubbles begin to form, check the underside often, cooking until slightly golden, and then flip crepe over. Check underside again until slightly golden.

Once thoroughly cooked, place cooked crepes on a plate. Keep separated with paper towel. Wash spatula clean and spray pan between batches as necessary.

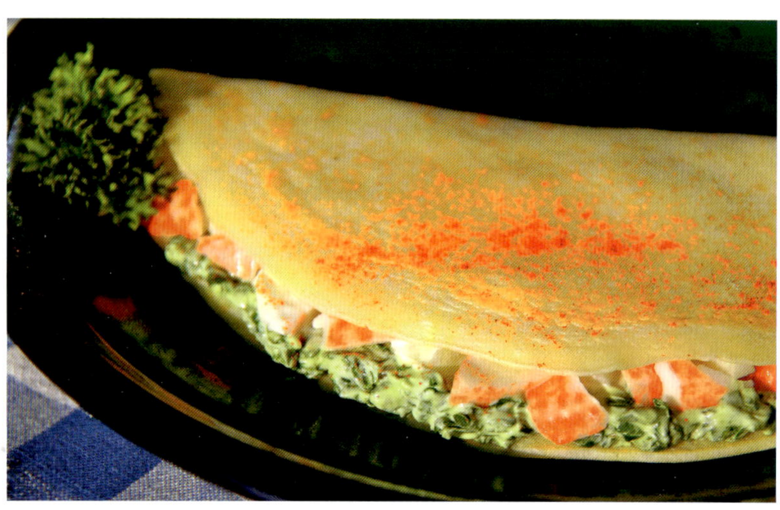

When crepes are ready, place the best-looking side face down. Spread half of each crepe with avocado filling first then top with lobster filling. Fold the other half of the crepe over and sprinkle with paprika.

Crepes and lobster filling can be made a day or two in advance, but the avocado filling should be prepared near serving time. Lobster and Avocado Crepes are best enjoyed at room temperature.

LOBSTER DIP

serves 4–6 as an appetizer or snack

The flavour and versatility of this dip will raise eyebrows. It can even be used as a condiment for vegetable dishes (mashed potatoes, broccoli, asparagus), to top salads, or to make little party sandwiches. With simple ingredients and easy instructions, this dip can be made in minutes.

- 2 cups chopped lobster meat
- 3/4 cup cheddar cheese
- 1 (8-ounce) block light cream cheese
- 1 cup light sour cream
- 1 teaspoon Worcestershire sauce
- 1/2 teaspoon Tabasco sauce
- 1/2 teaspoon finely grated garlic
- salt and pepper to taste

In a large bowl, blend room-temperature cheddar, cream cheese, and sour cream together with all other ingredients except lobster meat. Once mixture is smooth, add in the lobster meat. Refrigerate.

When ready to serve, heat lobster dip in microwave oven about 5 minutes or until simmering. Stir and heat again for 2 minutes. Serve at room temperature with your favourite bread, crackers, or veggie sticks.

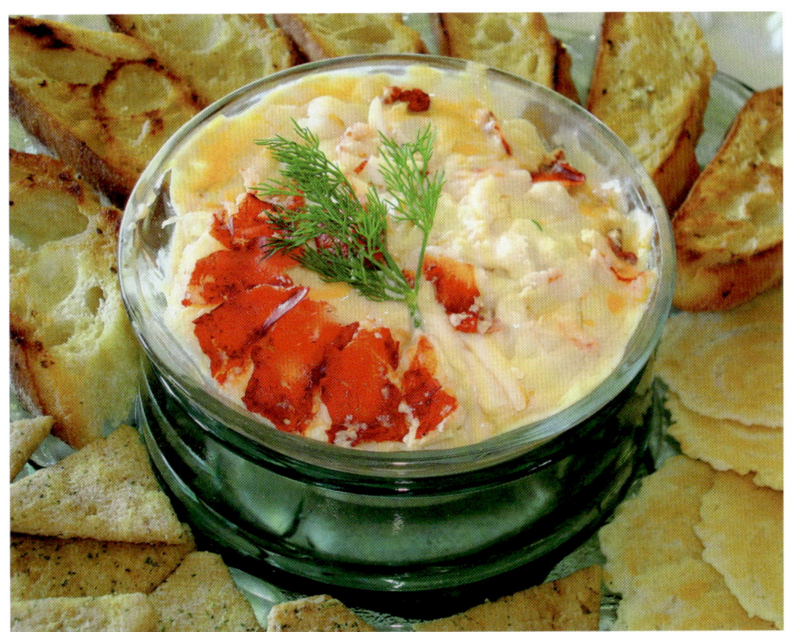

Lobster Side Dishes

When having lobster, some folks *just have lobster.* Perhaps they aren't sure what goes well with whole lobster in the shell. The Canadian Food Guide places a lot of emphasis on the importance of eating balanced meals for optimal health. With that in mind, here are five great easy side dishes that will enhance your lobster experience while introducing nutrients from other important food groups. For a complete, healthy meal, choose a grain-based dish and a vegetable side dish from these suggested recipes to accompany your succulent, protein-rich meal of Atlantic-caught lobster.

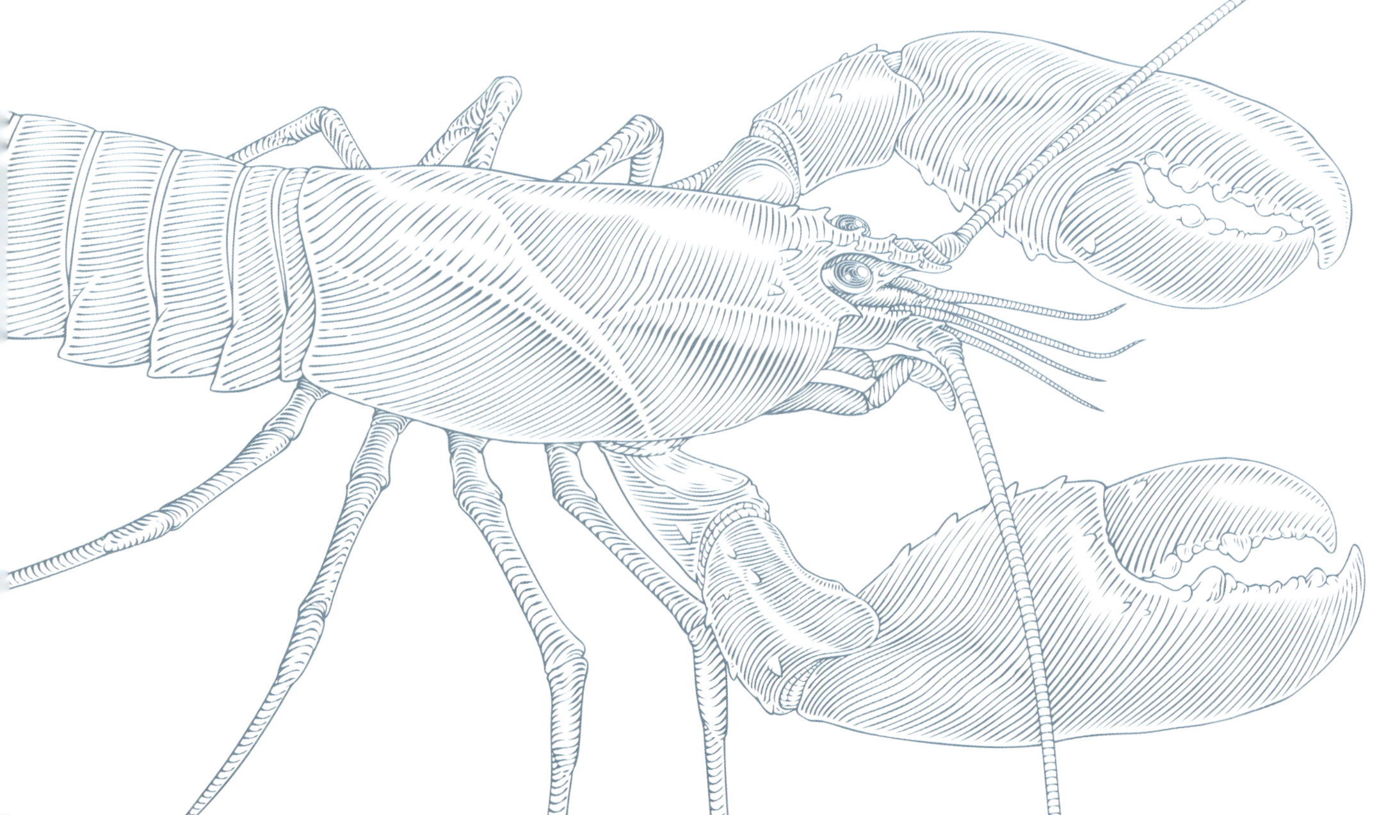

DILL POTATO SALAD

serves 4

The tangy flavour of fresh or pickled dill would give any traditional potato salad interest and originality—but this recipe is a well-tested favourite. A lovely, delicate accompaniment to juicy, briny whole lobster in the shell or with shelled lobster parts. The flavours contrast amazingly well.

- 6 medium white potatoes
- 4–5 slices cucumber dill pickles (sandwich–stacker-style)
- 1/2 cup mayonnaise (or to taste)
- 1/4 cup chopped chives
- 1 bundle fresh dill

Peel and boil potatoes until cooked but not too soft, about 20 minutes. Drain water, replace lid, and allow potatoes to cool. Dice potatoes once they've reached room temperature. Cut pickle slices into tiny cubes and add to diced potatoes. Add chives, 1/4 cup chopped dill, and mayonnaise. Mix well.

Serve on a bed of lettuce or fresh baby spinach leaves. Garnish with chopped dill.

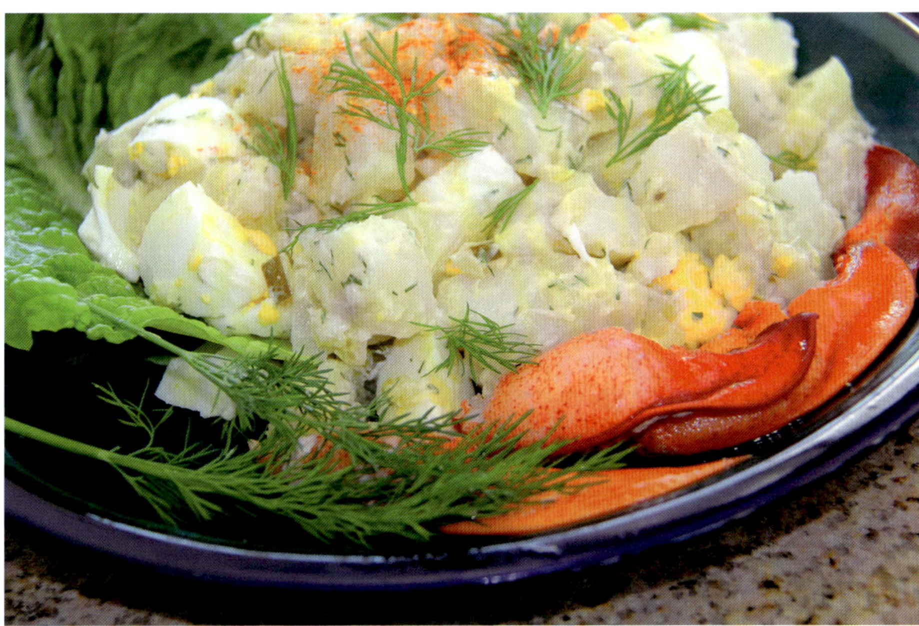

TANGY RED BELL PEPPER PASTA SALAD

serves 4

There is lots of red in this pasta recipe. Not only is it visually in harmony with whole red lobster in the shell, it is light, flavourful, and easy to prepare. Enjoy it in a large glass bowl to showcase the reds and allow guests to serve themselves: this dish is greatly enhanced when lobster juices on the plate accidently slip in.

- 1 (454g) box penne, gemeli, or orzo pasta
- 1 large clove garlic, grated
- 1 tablespoon sea salt
- 1 large red bell pepper, chopped
- 1/4 cup extra virgin olive oil
- 1/4 cup balsamic vinegar
- 3 tablespoons honey
- 1 teaspoon red pepper flakes (or to taste)
- 1/4 cup minced fresh parsley
- 2–3 thinly sliced radishes (optional)

Boil pasta el dente, according to directions. Meanwhile, place garlic, sea salt, chopped bell pepper, olive oil, balsamic vinegar, and honey in a blender or food processor and mince until pepper is reduced to confetti-size bits. Pour mixture into a large mixing bowl. Drain cooked pasta and add to mixture while still hot.

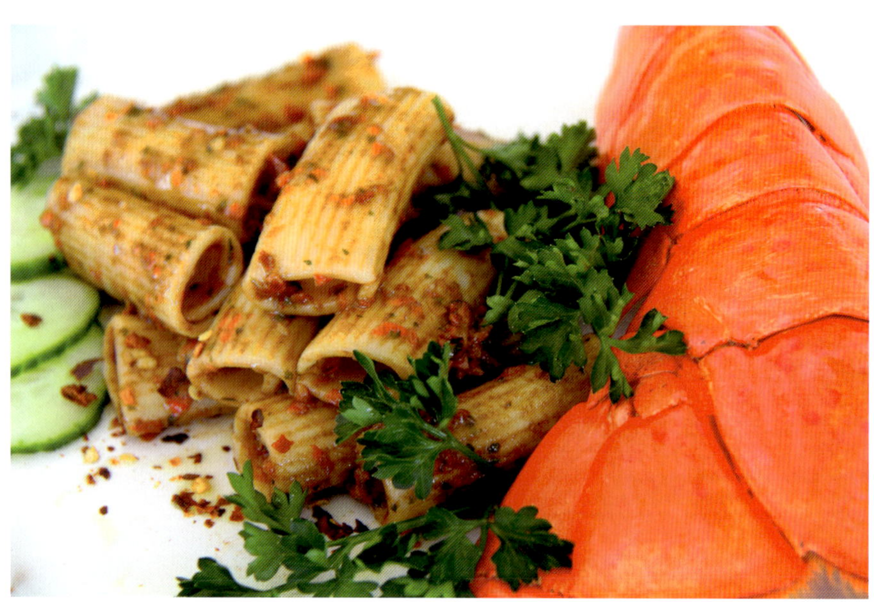

Toss with red pepper flakes and parsley, and garnish with optional radish slices for a light crunch. Cover for flavours to blend. Serve at room temperature.

FETA BROCCOLI SALAD

serves 4

This slightly crunchy vegetable side dish gives balance to a meal of whole lobster in the shell by introducing brilliant green parboiled broccoli and two additional colourful vegetables, including a dairy item. Throw in whole wheat buns or slices of artisan grain bread and you've got a well-balanced, nutritious meal.

- 4 heads broccoli
- 1 teaspoon sugar
- 1/4 cup red onion, finely chopped
- 1/3 cup mayonnaise
- 1/2 cup crumbled feta cheese
- 1/2 cup finely shredded blue cabbage
- 1/4 cup roasted sunflower seeds (optional)

Bring a pot of water to a rolling boil. Chop broccoli into bite-size florets. Blanch broccoli florets by dropping them in the boiling water for just 10 seconds, or until they are evenly bright green. Drain water and allow broccoli to cool.

In a large bowl, mix sugar, red onion, mayonnaise, and feta. Toss cooled broccoli into mixture. Cover in plastic and refrigerate.

When ready to serve, mix in blue cabbage (this way it won't cause the feta cheese to turn purple). Top each serving with sunflower seeds for an extra nutty crunch.

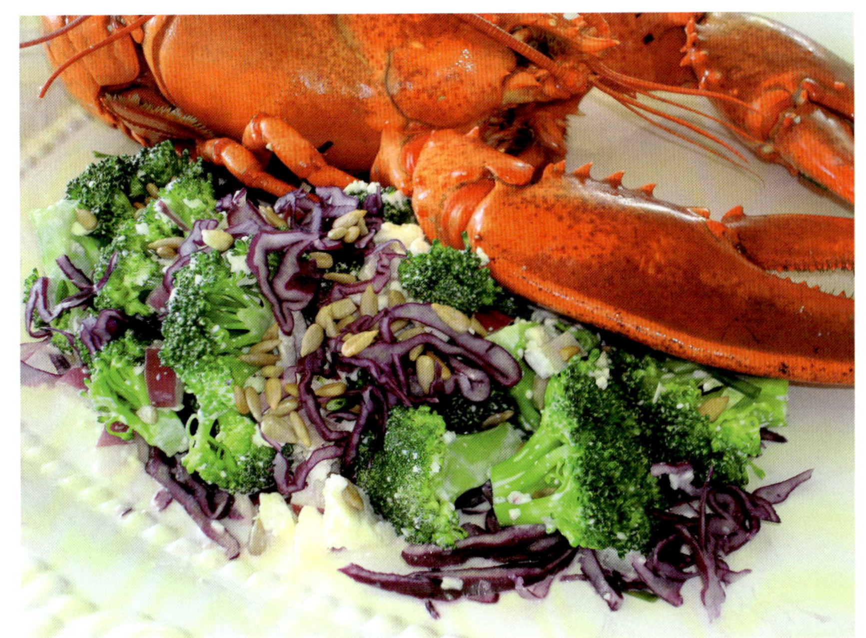

THAI MANGO SALAD

serves 4

Surprise your guests with a side of fruit. This sweet and tangy Thai salad is an absolutely wonderful accompaniment to a meal of briny lobster. The colours amber and red on light green alongside lobster in the shell make a gorgeous and appetizing presentation.

- 1 firm mango
- 1/2 sweet red bell pepper
- 2–3 chive sprigs
- 4 leaves iceberg lettuce
- 1 tablespoon lime juice
- 2 tablespoons sugar
- 2 tablespoons fish sauce
- 1 tablespoon oil (sunflower, canola, grape, or flaxseed)
- sprinkle finely chopped parsley and/or mint

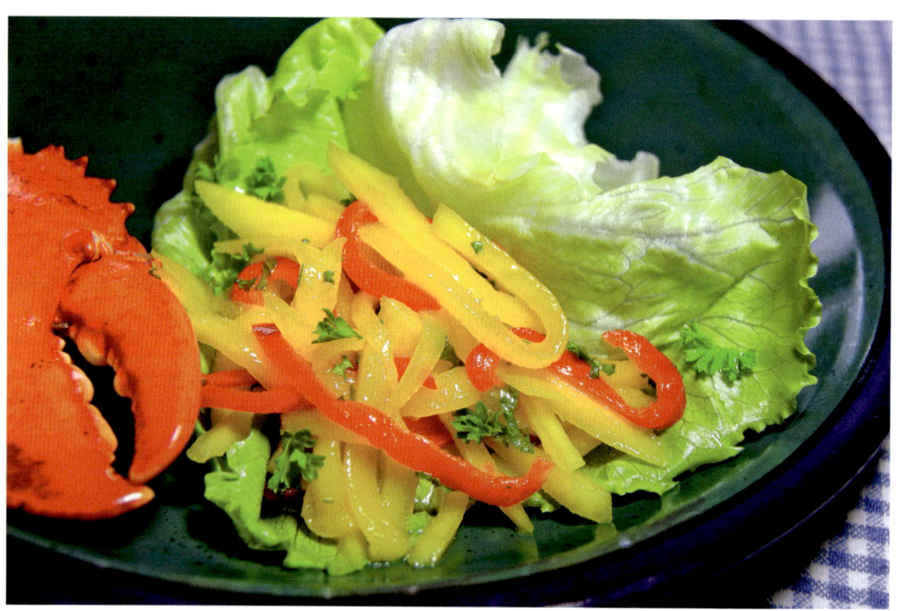

Peel mango and slice thinly, julienne style. Do the same with bell pepper. Using scissors, cut chive sprigs. Toss together with all other ingredients.

Divide into four portions and serve on top slightly curved crisp leaves of iceberg lettuce. Salad can be made one day in advance.

SWEET POTATO FETA CASSEROLE

serves 4–6

This is an amazing, flavourful side dish. Sweet potatoes have more nutritious value and are lower in carbohydrates than regular potatoes. Furthermore, their orange-red flesh is very attractive on any plate but especially with whole lobster in the shell. The crumbled feta cheese will not melt when heated, but will in your mouth.

- 2 large sweet potatoes
- 3 tablespoons olive oil
- freshly ground pepper
- 1/2 cup crumbled feta cheese
- 1/3 cup pine nuts
- drizzle lemon juice (optional)

Preheat oven to 400°F (200°C). Peel and cut potatoes into 1-centimetre slices. Place cut potatoes in a baking dish large enough to spread them in a single layer. Add oil and pepper. Toss well, then spread slices. Cover the dish with aluminum foil and bake for about 20 minutes.

Remove foil, sprinkle feta and pine nuts evenly, and continue to bake uncovered another 8–10 minutes or until potatoes slices are tender. Serve with a drizzle of lemon juice for extra flavour.

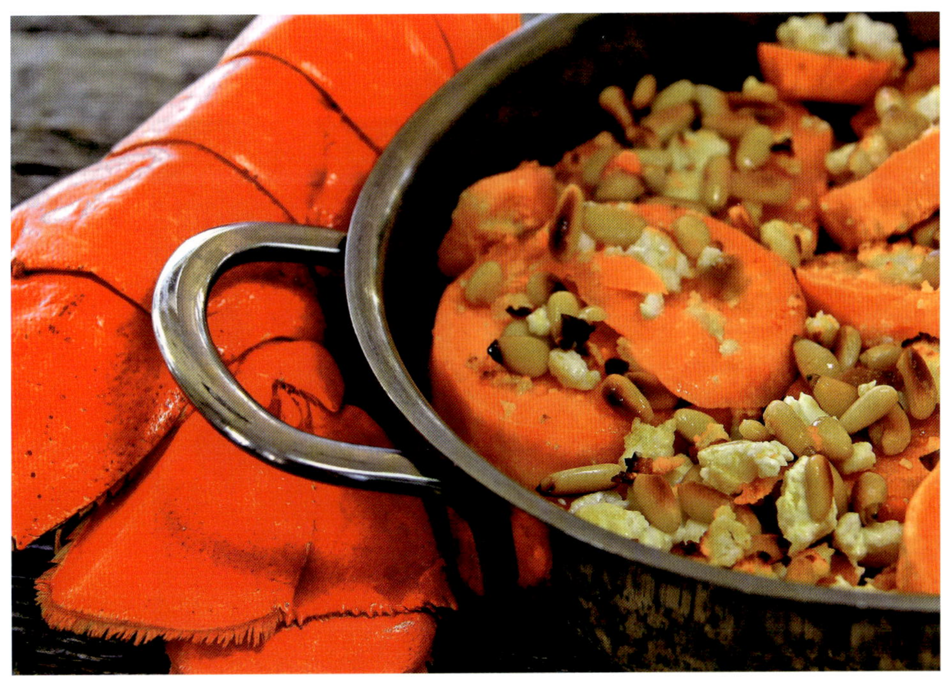

GARLIC STEAMED RICE

serves 4

As delicious as lobster is on its own, it's always a good idea to have a starchy side dish. This rice recipe is a nice change from potato salad. It is surprisingly fluffy and the lemon-garlic undertones compliment seafood, in particular lobster, very well. Spoon it onto the plate next to lobster with a fresh sprig of your favorite herb. Delightful!

- 1 1/2 cups rice (white, long-grain, basmati, or jasmine)
- 1 large clove garlic, peeled (add more for extreme garlic lovers)
- 1 teaspoon sea salt
- 1/2 cup extra-virgin olive oil
- 3 tablespoons Dijonnaise mustard
- 1 teaspoon lemon juice

Steam rice according to directions. Once cooked, place in a large mixing bowl. Grate garlic onto a plate. Using a fork, crush garlic into sea salt. Add olive oil. Mix in mustard and lemon juice. Pour over prepared rice then stir it in. Cover with plastic at room temperature for flavours to blend with rice. Heat in microwave just before serving.

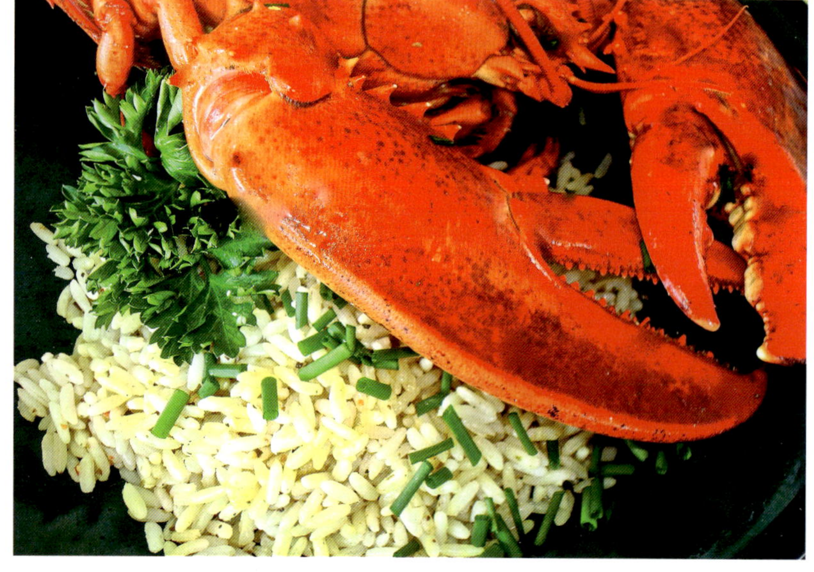

FESTIVE SPRING FIDDLEHEADS

serves 4

This little wild vegetable is the perfect accompaniment for spring-caught lobster. Many Acadian families celebrate the arrival of spring with a feast of lobster and fiddleheads—the ultimate "eat local" meal. One of the first plants to pierce the ground before the final spring thaw, fiddleheads inhabit loamy damp soils of inland riverside forests. Harvesters rarely share where their favourite patches are, but fiddleheads are readily available in Maritime supermarkets from early to late June. Fiddleheads freeze well once parboiled (2 minutes in salted boiling water) and sealed in plastic bags to enjoy year-round. They have a unique nutty or tea-like flavour, and a layered and sublimely creamy texture.

- 4 generous handfuls of fresh fiddleheads
- 1 tablespoon butter or virgin olive oil
- Sea salt and cracked pepper to taste

Bring a large soup pot of water to a boil. Meanwhile, rinse fiddleheads thoroughly until water is clear. Pick through and remove as many rust-coloured papery flakes as possible. Clip off dark stem-ends. (Stems shouldn't be any more than 5 centimetres long.) Boil for about 20 minutes or until tender to the bite. Drain and serve piping hot with butter or a drizzle of olive oil, salt, and pepper to taste.

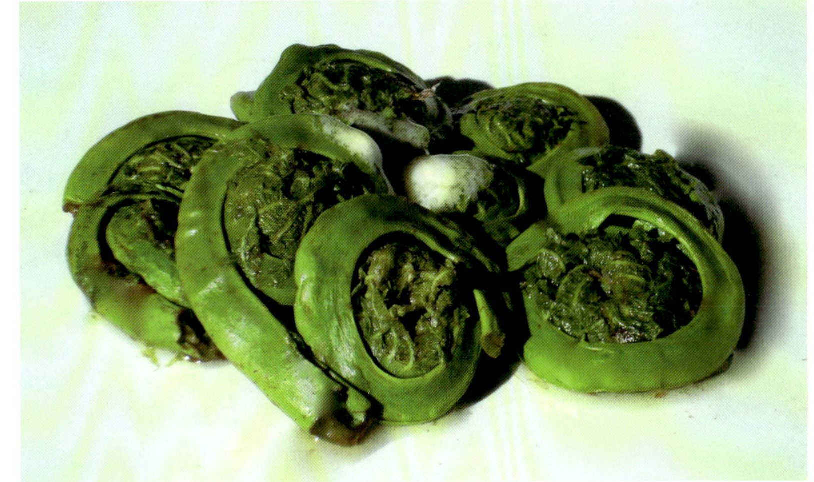

QUIZ ANSWERS

Multiple Choice

1. True
2. False
3. False
4. False
5. True
6. True
7. True
8. True
9. D
10. B
11. B
12. C
13. D
14. D
15. B

Short Answer

16. A lobster's greatest predator is YOU—once it reaches market size. Lobster larvae and juveniles fall prey to plankton eaters, fish, seabirds, and seals. But that's fair game. Fishers, on the other hand, trick lobsters into a trap with the lure of bait.
17. So-called clawless, or spiny, lobster are not technically lobster at all. Although similar in appearance to Atlantic Canada's cold-water lobster, these are two different groups of crustaceans that are not closely related. The clawless "lobster" of warmer waters is more like a large shrimp with long, spiny antennae. The only edible part of the spiny lobster is its tail, which tends to be tough and less succulent than North Atlantic lobster.
18. Soft-shell lobster are less meaty and less tasty. In the first week following a moult, before the new shell hardens as it fills with water, the lobster experiences a rapid growth spurt. Consumers get a much better dining experience and better value out of a hard-shell lobster, as the meat inside is beefier. Furthermore, soft-shell lobster are less hardy and not as likely to survive the trip to market.

19. In their early life stages, lobster are swimmer-drifters. Once they grow to what we know to be the "lobster" with claws, they become bottom dwellers and crawl to get around. However, if startled, lobster will swim backwards by thrashing their tail inward and out. This enables them to reach a speed of 5 metres per second (11 mph).
20. Yes, there are parts of the lobster that can make you sick: don't eat the shells! Unless you have an allergy to shellfish, that really is the only part of a lobster your digestive tract cannot handle. As a rule, people don't find lobster eyeballs and antennae appetizing, but enjoy what appeals to you.

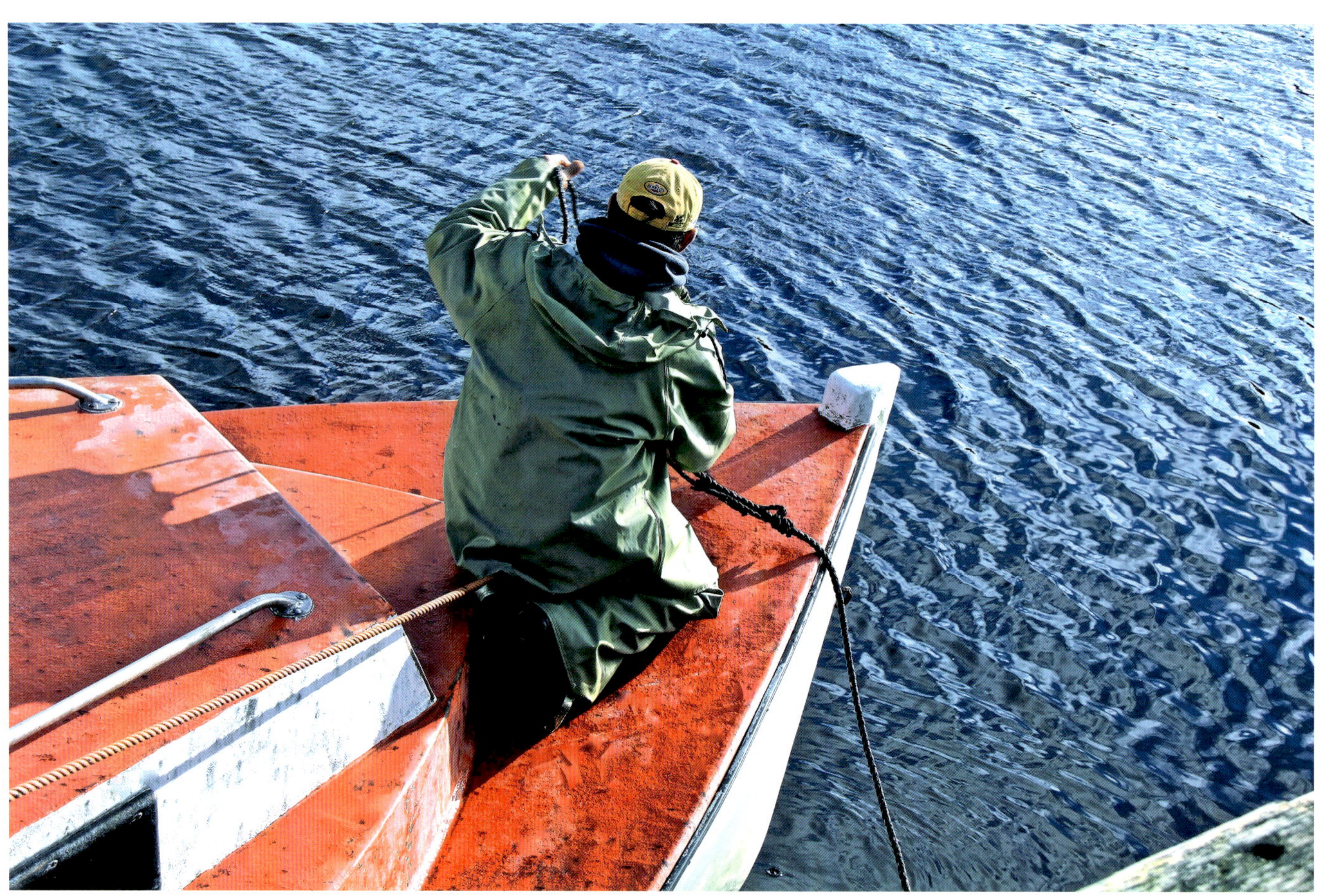

Tying up after a hard day of lobstering.

RESOURCES

Books and Articles

Bavington, Dean. *Managed Annihilation: An Unnatural History of the Newfoundland Cod Fishery.* Vancouver: UBC Press, 2010.

Corson, Trevor. *The Secret Life of Lobsters: How Fishermen and Scientists are Unraveling the Mysteries of Our Favorite Crustacean.* New York: HarperCollins, 2004.

Deryshire, David. "Do Lobsters Hold the Key to Eternal Life?" Daily Mail Online, September 2013. Dailymail.co.uk

Jenson, L. B. *Fishermen of Nova Scotia.* Halifax: Petheric Press/Nimbus Publishing, 1988.

Kerr, W. P. *Port-Royal Habitation: The Story of the French and Mi'kmaq at Port-Royal,* Halifax: Nimbus Publishing, 2005.

King, Richard J. *Lobster.* London, UK: Reaktion Books, 2011.

Lemieux Oragano, Christina. *How to Catch a Lobster in Down East Maine.* Mount Pleasant, South Carolina: Arcadia Publishing, 2012.

Palmer, Roxanne. "Maine Lobster Threatened by Climate Change, Scientist and Industry Warn." *International Business Times,* US, 2013.

Reiley, Amy. "Lobster—the Luxury Source of Protein" Eatsomethingsexy.com, n.d.

Townsend, Elisabeth. *Lobster: A Global History.* London, UK: Reakton Books, 2011.

Withrow, Alfreda. *St. Margaret's Bay: A History.* Glen Margaret, NS: Four East Publications, 1985.

Websites

Canadian Council of Professional Fish Harvesters: fishharvesterspecheurs.ca

Department of Fisheries and Oceans: dfo-mpo.gc.ca

Lobster Anywhere: lobsteranywhere.com

Lobster Council of Canada: lobstercouncilcanada.ca

MarineBio: marinebio.net

Oceanic Research Group: oceanicresearch.org

Pictou-Antigonish Regional Library: parl.ns.ca/lobster

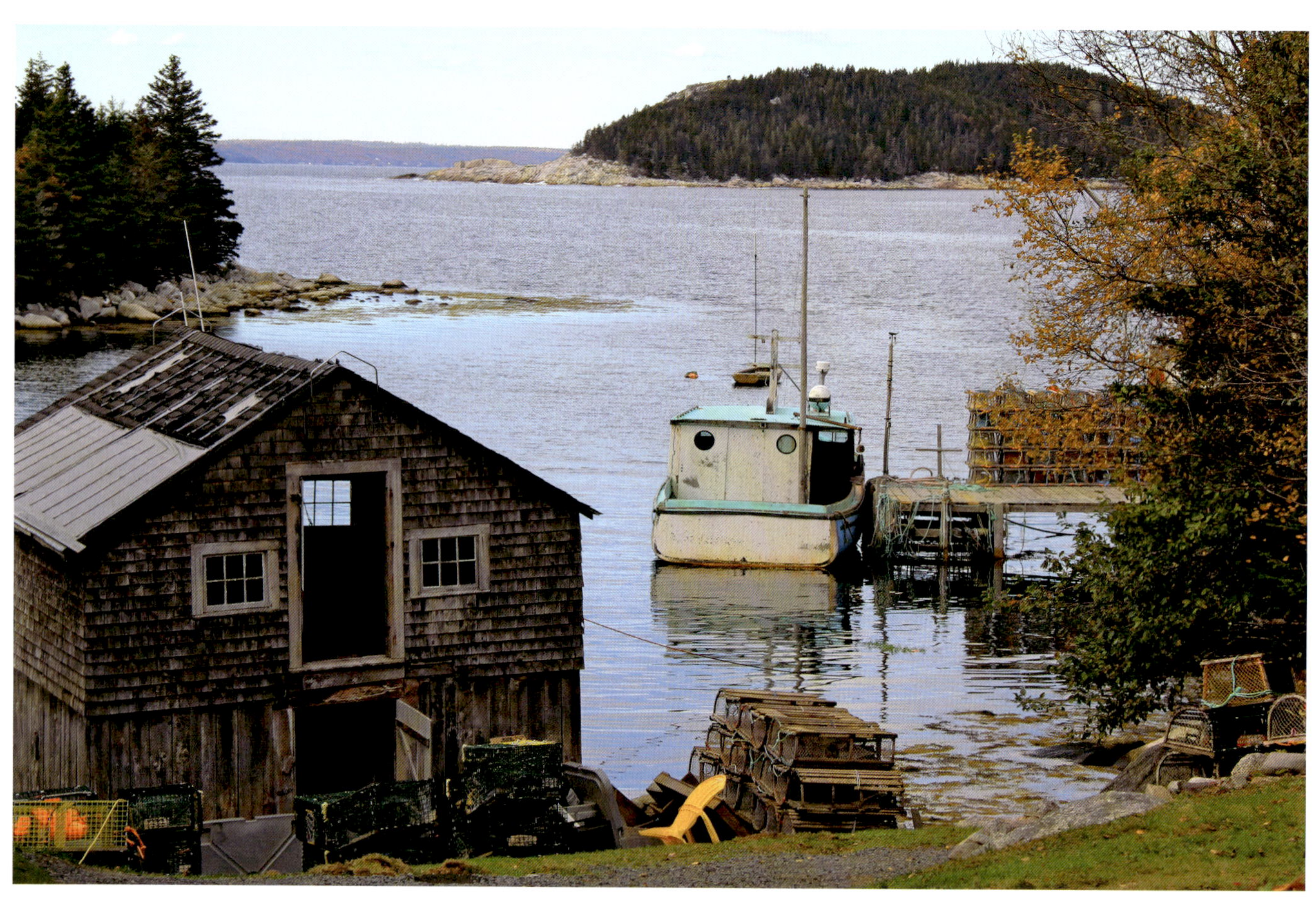

View from Modesty Cove, St. Margarets Bay, Nova Scotia.

IMAGE CREDITS

All images by Denise Adams, except where noted below.

Canstock

p. 16 (top): European Lobster - British Isles (csp9358656)

p. 16 (bottom): Common slipper lobster (csp2163363)

p. 17 (top): Caribbean Spiny Lobster (csp0690678)

Jenn Embree

p. 90: DFO fishing zones map

Frank Gale for the *Gulf News*

p. 42 (bottom left): published July 1, 2013

Nova Scotia Archives

p. 48: Nova Scotia Information Service Nova Scotia Archives no. NSIS 10494

p. 49: *Nova Scotia's Ocean Playground*, 1960; NSA Library Collection: F91 N85 N85

p. 50: Alexander H. Leighton Nova Scotia Archives accession no. 1988-413 / negative no.: 707-d

p. 58 (top): *Packing lobster in factory near Cheticamp*, Nova Scotia Information Service, photographer, c. 1959; NSA, NSIS Photo no. 13393

p. 88: *Hauling Lobster Traps*, William E. deGarthe, artist, 1971; NSA Documentary Art Collection 1986–352 no. 15

Peggy Kochanoff

p. 13: Lobster cross-section

p. 34: Lobster life cycle

Scott Leslie

p. 15: American lobster

Shutterstock

p. 22 (top): *Nascita di Venere.* The birth of Venus (Botticelli) (80044615)

p. 40: Micrasterias truncata algae (136145006)

Tin-Yam Chan, National Taiwan Ocean University

p. 17 (bottom): Japanese fan lobster

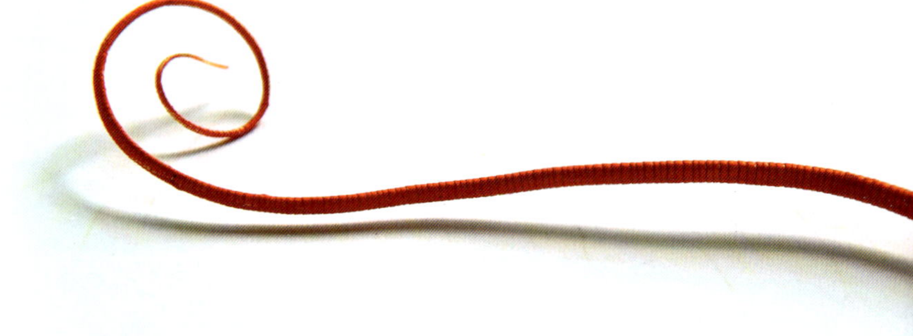